W9-AAE-149

Back to Basics

Your Guide to Manufacturing Excellence

The St. Lucie Press/APICS Series on Resource Management

Titles in the Series

Applying Manufacturing Execution Systems
by Michael McClellan

Back to Basics:
Your Guide to Manufacturing Excellence
By Steven A. Melnyk
R.T. "Chris" Christensen

Enterprise Resources Planning and Beyond:
Integrating Your Entire Organization
by Gary A. Langenwalter

ERP: Tools, Techniques, and Applications
for Integrating the Supply Chain
by Carol A. Ptak with Eli Schragenheim

Integral Logistics Management:
Planning and Control of Comprehensive Business Processes
by Paul Schönsleben

Inventory Classification Innovation:
Paving the Way for Electronic Commerce and
Vendor Managed Inventory
by Russell G. Broeckelmann

Macrologistics Management:
A Catalyst for Organizational Change
by Martin Stein and Frank Voehl

Restructuring the Manufacturing Process:
Applying the Matrix Method
by Gideon Halevi

Supply Chain Management:
The Basics and Beyond
by William C. Copacino

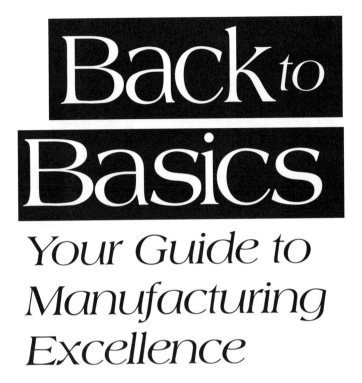

Back to Basics

Your Guide to Manufacturing Excellence

STEVEN A. MELNYK

AND

R.T. "CHRIS" CHRISTENSEN

The St. Lucie Press/APICS Series on Resource Management

S^t_L

St. Lucie Press
Boca Raton • London
New York • Washington, D.C.

**THE EDUCATIONAL SOCIETY
FOR RESOURCE MANAGEMENT**
Alexandria, Virginia

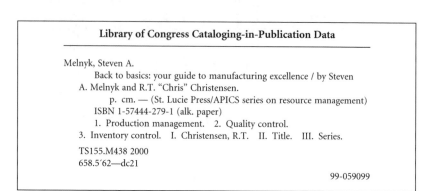

Library of Congress Cataloging-in-Publication Data

Melnyk, Steven A.
    Back to basics: your guide to manufacturing excellence / by Steven
A. Melnyk and R.T. "Chris" Christensen.
        p. cm. — (St. Lucie Press/APICS series on resource management)
    ISBN 1-57444-279-1 (alk. paper)
    1. Production management.   2. Quality control.
3. Inventory control.   I. Christensen, R.T.   II. Title.   III. Series.
    TS155.M438 2000
    658.5′62—dc21
                                                                99-059099

© 2000 by CRC Press LLC
St. Lucie Press is an imprint of CRC Press LLC

No claim to original U.S. Government works
International Standard Book Number 1-57444-279-1
Library of Congress Card Number 99-059099
Printed in the United States of America 1 2 3 4 5 6 7 8 9 0
Printed on acid-free paper

The Authors

 **Steven A. Melnyk,** CPIM, is Professor of Operations Management at Michigan State University. His research interests include MRPII, supply chain management, metrics/system measurement, time-based competition (TBC), and environmentally responsible manufacturing (ERM). Steve is the lead author of six books, *Shop Floor Control* (Dow Jones-Irwin, 1985), *Production Activity Control: A Practical Guide* (Dow Jones-Irwin, 1987), *Shop Floor Control: Principles, Practices, and Case Studies* (APICS, 1987), *Computer Integrated Manufacturing: A Source Book* (APICS, 1990), *Computer Integrated Manufacturing: Guidelines and Applications from Industrial Leaders* (Business One, 1992), and *Operations Management: A Value Driven Perspective* (Irwin Press, 1996). He was also the editor of the first work devoted entirely to the study of tooling in production and inventory control (Grand Rapids APICS/Michigan State University, 1988).

Steve's articles have appeared in such international and national journals as the *International Journal of Production Research, International Journal of Operations and Production Research, Journal of Operations Management, International Journal of Purchasing and Materials Management,* and *Production and Inventory Management.* Along with Dr. Handfield, he was co-editor of a special issue of the *Journal of Operations Management* that dealt with theory-driven empirical research. In 1995, Steve received a major grant from the National Science Foundation to study environmentally conscious manufacturing (a joint grant between the Colleges of Business and Engineering at Michigan State University). In addition, he sits on the editorial review board for *Production and Inventory Management* and the *International Journal of Production Research.* Steve is also the software editor for *APICS: The Performance Advantage,* as well as the lead author on the monthly "Back to Basics" column found in that magazine.

Steve is a member of the Mid-Michigan Chapter of APICS and a past Director of Publicity for Region XIV of APICS. He is also a member of the Central Michigan Chapter of NAPM. Steve has spoken to various APICS and NAPM groups at the international, national, and regional levels and was also a member of the Delivery of Products and Services committee of the Certification in Integrated Resource Management (CIRM) examination program developed by APICS. He has consulted with over 60 companies, including the American Red Cross, Tokheim, and General Motors. He has also taught manufacturing principles to universities and firms located in Hungary, Poland, and Czechoslovakia as part of a consortium program developed by Midwestern Universities. His contributions to the theory and practice of production and inventory management were recognized in 1992 by the Grand Rapids Chapter of APICS when they awarded him the Paul Berkobile Award. In addition, in 1999, he was recognized as a Faculty Pioneer by ASPEN Institute for his leading-edge work on integrating environmental concerns into business practice.

Steve is listed in the 1995 to 1999 issues of *Business Week: Guide to the Best Business Schools* as one of the ten outstanding faculty members of Michigan State University's MBA Program. He has also taught in Michigan State University's prestigious Advanced Management Program. Finally, he is in *Who's Who* at the Midwest, National for Education, National, and International levels.

R.T. "Chris" Christensen is the Director of Operations Management programs at the University of Wisconsin-Madison, School of Business, Executive Education. He is responsible for all seminars and workshops dealing with manufacturing management and maintenance management at the Institute. Chris has over 25 years of industry experience in progressively responsible line, staff, managerial, and executive positions in diverse manufacturing environments. His corporate experience includes positions with Tenneco's Walker Manufacturing, Johnson Controls, Allis-Chalmers, and Harnischfeger Industries. Since joining Management Institute in 1988, Chris has enhanced the production/operations program area by developing timely state-of-the-art seminars that address current issues and provide solutions to some of today's most challenging problems in manufacturing. He developed and teaches in the Maintenance Management Certificate Series, the first university-sponsored maintenance series offered in the U.S. He also provides consulting services and in-house training to companies expanding their horizons and implementing new manufacturing controls to address millennium 2000 challenges.

Chris is very active with professional and trade associations. He is a member of APICS (American Production and Inventory Control Society) and currently serves as chair of APICS CIRM (Certification in Integrated Resource Management) module for DP&S (Designing Products and Services). He is a member of WERC (Warehousing Education and Research Council), senior member of SME (Society of Manufacturing Engineers), past board member of CMED (Conference on Management and Executive Development), and past vice-president education for APICS-Madison chapter. He serves on several university committees, including the Wisconsin Extension Education Advisory Board and is an elected representative to the statewide University Extension Council. He is also an advisory board member of a regional food corporation.

Chris has authored numerous articles in professional association publications. He is co-author of the monthly column "Back to Basics" in the APICS magazine and is a subject matter expert and co-author of APICS DP&S training materials. He is a frequent speaker for many professional organizations and University of Wisconsin groups. Chris holds a Bachelor of Business Administration degree in Manufacturing Management and an MBA in Manufacturing from the University of Wisconsin.

Acknowledgments

This book (and the series on which it is based) is the result of the contributions of several people and groups. We would like to acknowledge their contributions by thanking them publicly. We would like to extend our appreciation to the American Production and Inventory Control Society (APICS), Phil Carter of the Center for Applied Purchasing Studies (CAPS) of the National Association of Purchasing Management, Jerry Bapst, William R. Wassweiler, Don Franks, Ben Schussel, William (Bill) Jones, Nat Natarajan, John Llewellyn, Carol Ptak, Richard (Dick) T. Smith, Gene Woolsey, Tim Hinds of Michigan State University, and Randall Schaefer. We would also like to thank the numerous readers of the column who have come to us with their comments and suggestions for future topics. They have helped to develop and shape the column and this book.

Preface

The "Back to Basics" concept can be traced back to the 1996 APICS International Conference that took place in New Orleans, LA. At that time, one of the authors, Steven A. Melnyk, was asked to do two presentations. One of these presentations was on the advanced manufacturing topic of environmentally responsible manufacturing, while the other was on the basics of the Internet for production and inventory control personnel. The first topic, scheduled for a room designed to hold over 400 people, was attended by less than 20 people. The other, more basic topic was scheduled for the same room. It was packed to overflowing. This difference in attendance was noted and shared with several other speakers, one of whom, Bill Wassweiler, observed that the presentations that dealt with basics tended to generate the greatest amount of attention and interest. However, these types of presentations were often seen as being too simple and elementary — everyone should know the basics. This discussion was continued with the other author of this book, R.T. "Chris" Christensen, who had noticed that questions of a basic nature become key elements in discussions he has with business people in his executive seminars held at the University of Wisconsin.

From these observations, a paradox began to emerge. The manufacturing world was being shaped by such new developments as Enterprise Resource Planning (ERP), computer simulation, and advanced planning systems. Yet, these topics, while important, did not seem to generate the same high level of enthusiasm as we were seeing for topics dealing with such basic concepts as part numbering, capacity, processes, inventory accuracy, and bills of materials. This paradox was explored with people attending the conference and others. What began to emerge was an interesting picture of the knowledge base and of the need for a series of articles dealing with the basics and their mastery.

Production and inventory control is an interesting field. It is an entry point for many people as they enter the world of operations management, when they are typically assigned to areas such as planning, scheduling, or inventory control. When they enter these fields, most find that they are challenged by the need to know and master the basics; however, mastering such basics is no simple task. In many cases, it requires that they take extra courses (offered by organizations such as APICS or by colleges or universities) or that they read extensively on their own. Developing this mastery takes time, and often this mastery comes through the "school of hard knocks" — mistakes are made and people learn from their mistakes. This is not a very effective and efficient way of learning. After a period of time, though, people do develop a mastery of the basics, and at that point they are better able to appreciate the impact of such new procedures as ERP or Collaborative Planning and Forecast Replenishment (CPFR) or Vendor Managed Inventories (VMI). As soon as people get to this point, they are most likely to be promoted, and their replacements are very much like they were when they first entered the field — and the cycle is repeated.

Something had to be done to break this cycle. There has been a great deal of confusion about the basics. People want to have access to very focused discussions on various elements comprising the basics, and providing such exposure would be the major goal of a "Back to Basics" series. This was the concept that was proposed to the publisher of the journal, *APICS: The Performance Advantage*, who proposed a test run beginning in January 1997. The initial response was enthusiastic, to say the least. From that response, a column was born that has continued to grow until it is now one of the two most widely read columns in the journal.

Premises Underlying "Back to Basics" Approach

This series of columns is based on a number of important and critical premises. The first is that the columns have to be rigorous in their treatment of the basics but also management oriented. This is the reason why the column is co-authored by Steven A. Melnyk of Michigan State University and R.T. "Chris" Christensen of the University of Wisconsin at Madison. Steven Melnyk is primarily a researcher and teacher in the operations management area. His task in the process is to ensure that the treatments are accurate, rigorous, and sufficiently comprehensive. Complementing him is Chris Christensen, who brings over 25 years of practical, real-world experience in manufacturing to his current position in executive education. His role is to make sure that the treatments of the topics are relevant to management.

The second premise underlying the series is that the authors will deal only with the basics. This is not the place to look for information about such new developments as ERP, advanced planning systems, or CPFR. Rather, this is the place to find information about such issues as capacity, inventory, inventory accuracy, part numbering, processes, problem solving, and general managerial guidelines.

The third premise is that mastery of the basics is critical to any success in operations management. One cannot be expected to succeed in the complexity of operations management without an understanding of the basics. Further, this treatment of the basics should cover not only the basics but also the subtleties of the basics.

The fourth premise is that these basics are applicable to anyone working in operations management. They can be used by people working in manufacturing or in hospitals or in a bank. The context may be different, but the basics remain the same.

Finally, it was assumed that exposure to the basics should be managed through the use of "bite-size" treatments. That is, each treatment should be information dense and not require that the reader devote a great deal of time and effort. This last premise has shaped the resulting structure of the articles.

Structure of the Articles

Each section found in this book (and each "Back to Basics" column) is designed to run between 1000 and 1500 words. This length, it has been found, is just about right. It allows sufficient development of the topic being discussed but is short enough to be read quickly. Such a structure forces the articles to deal with the basics in the most direct way. There is little room for "fluff." Further, each article is designed to be self-contained. That is, when you finish reading an article, you have been introduced to a set of critical issues and concepts. Each article has everything that you need to understand the issues and to appreciate their implications.

About APICS

APICS, the Educational Society for Resource Management, is an international, not-for-profit organization offering a full range of programs and materials focusing on individual and organizational education, standards of excellence, and integrated resource management topics. These resources, developed under the direction of integrated resource management experts, are available at local, regional, and national levels. Since 1957, hundreds of thousands of professionals have relied on APICS as a source for educational products and services.

- **APICS Certification Programs** — APICS offers two internationally recognized certification programs, Certified in Production and Inventory Management (CPIM) and Certified in Integrated Resource Management (CIRM), known around the world as standards of professional competence in business and manufacturing.
- **APICS Educational Materials Catalog** — This catalog contains books, courseware, proceedings, reprints, training materials, and videos developed by industry experts and available to members at a discount.
- *APICS: The Performance Advantage* — This monthly, four-color magazine addresses the educational and resource management needs of manufacturing professionals.
- **APICS Business Outlook Index** — Designed to take economic analysis a step beyond current surveys, the index is a monthly manufacturing-based survey report based on confidential production, sales, and inventory data from APICS-related companies.
- **Chapters** — APICS' more than 270 chapters provide leadership, learning, and networking opportunities at the local level.
- **Educational Opportunities** — Held around the country, the APICS International Conference and Exhibitions, workshops, and symposia offer numerous opportunities to learn from your peers and management experts.

- **Employment Referral Program** — A cost-effective way to reach a targeted network of resource management professionals, this program pairs qualified job candidates with interested companies.
- **SIGs** — These member groups develop specialized educational programs and resources for seven specific industry and interest areas.
- **Web Site** — The APICS web site at **http://www.apics.org** enables you to explore the wide range of information available on APICS' membership, certification, and educational offerings.
- **Member Services** — Members enjoy a dedicated inquiry service, insurance, a retirement plan, and more.

For more information on APICS programs, services, or membership, call APICS customer service at (800) 444-2742 or (703) 237-8344, or visit **http://www.apics.org** on the World Wide Web.

Contents

Dedications

Steven A. Melnyk
This book is dedicated to the following people who have influenced my development as a writer and as a teacher:

William R. Wassweiler (who taught me to understand the basics of shop floor control)

Randall Schaefer (who taught me to understand the practice of management)

Art Halstead (my father-in-law, who died on July 30, 1997 — a dear friend and thorough critic of the "big noise from Hamilton," his nickname for me)

Christine A. Melnyk (my wife, partner, and in-house editor — with this book, you have earned your bachelor in operations management)

Chris Melnyk (brother, friend, and confidant — a real collaborator in this book)

R.T. "Chris" Christensen
This book is dedicated to the following people who have influenced my development as a writer and as a teacher:

John J. Jepsen (who had sufficient faith in my manufacturing capabilities to give me the opportunities to learn)

William "Bill" J. Jones (who taught me the importance of interrelationships in manufacturing operations)

William "Bill" A. Wheeler (who taught me how to bring my work experiences to the classroom)

The Production/Operations Management Advisory Board, Executive Education, School of Business, the University of Wisconsin at Madison (this book is dedicated to the dozen business professionals who serve on the board and who give the seminars at which we present the focus and direction that this book brings forward)

Mary Jo Christensen (my wife and partner, who gave me the time to write this book and without whose help and understanding I would not have been able to complete this undertaking)

1 Why the Basics?

Understanding the Basics

In this, the first chapter of the book, we explore some of the foundations on which this book and the various topics discussed in it are based. It is here that we show how important the basics are, in that everything we do in operations management, in general, and in manufacturing, specifically, are built on them. We will also show that, when it comes to technology, it is best to work on the basics first, *before* seriously considering implementation of the latest and best in technology, be it hardware or software. In fact, we might consider the following hierarchy: (1) basics, (2) simplification — taking the basics and making them work better by simplifying them and making them easier to utilize, (3) integration — making sure that the basics in each area of the operations management/manufacturing system are working with each other rather than against each other, (4) focus — making sure that the basics we are implementing are consistent with the strategic stance of the firm, and (5) technology — after we have done everything that we can on the preceding four dimensions, we are ready and able to successfully implement technology and take advantage of it.

In other words, technology should always come after basics, not before, even though there are several success stories out in the field where the technology — often in the form of a new Material Requirements Planning (MRP)/ MRPII system — comes first. However, these implementations have served as catalysts for a reintroduction of the basics, and ultimately these implementations force management and the firm to return to the basics.

In this chapter, we turn our attention to three major issues:

1. Why the basics
2. The need to understand the map of manufacturing excellence
3. Understanding lead time

Oh, No! Not Another Book on Manufacturing!

Now that you have bought this book (or are at least reading this in the bookstore before deciding whether or not to purchase this book), you are asking yourself this question, "Why on earth do I need this book?" After all, this is a book on basics, and you know that you are far beyond the basic requirements of your operation. You may think to yourself that you are now working on the fifth generation of operating systems at your company and you see no need to get back to the basic stuff. There is no way that the information in this book could relate in any way to what you are doing. But it does.

How Can Things Be So Wrong, When Everything Seems To Be So Right?

Not that long ago, one of the authors began working with a corporation that seemed to have all the pieces in place to run a successful manufacturing operation. The company was a manufacturer of custom and low-volume repetitive special machinery. They had a scheduling system in place in engineering, were designing to schedule, and were quite successful in meeting the projected times in engineering. Some of these products were quite large, weighing several hundred tons when finished, and required up to six months of design, manufacturing, scheduling, and production time.

The company had about 250 hourly employees, a master scheduling system, and a shop floor dispatch system. If you looked at the system, as the author did as a consultant, everything had the appearance of being "fine," but it wasn't. On-time deliveries were incredibly late. Shop floor productivity was poor. Idle time was high for both man and machine. The author spent six months studying the operation from his position in the corporate office to find the solution to the problem.

He looked at the entire operation and talked to all the people in the firm from the president on down to the shop floor production operator, and one thing became clear. They were all doing their job and doing it correctly … well, correctly within acceptable tolerances in manufacturing, anyway. Nothing is perfect, and this was true here, but there was nothing that you could say was obviously out of line, and there was no one on the chain of responsibility from the top to the bottom who was doing his or her job in an unacceptable manner. The systems were in place. It was a MRP/MRPII shop with a robust master schedule and shop floor system. The system even appeared to be up to date and providing timely and correct data for the activities.

At the end of the month, though, the product managers were on the shop floor expediting their pet jobs. Shipments were late, and there was a major scramble on the shop floor at the end of the month to get products out the door. Chaos reigned, and costs went through the roof while fewer and fewer jobs made their due dates.

There was something fundamentally wrong here. If you think about it, there is no real logic in a system (or the activities driven by it) that places more value on products shipped at the end of the month than on any other date in the month, such as the third or seventh or twelfth day. If anyone can explain why we make more money by shipping at the end of the month, other than to play games with the numbers, please do so.

Anyway, everything seemed to be in place and looking good, but the company was, in fact, still finding itself in an end-of-the-month crunch. We would bet that a lot of you out there are in the same situation, even if you are too proud to admit it. The author just about went crazy trying to figure out what was happening, but then he found it. The answer was so basic that everyone had completely overlooked it. What was happening was that there was no visibility in the schedule. By that we mean that the engineering scheduling system was not visible to any other part of the system. The master schedule was tucked away under lock and key because it was "confidential." How absurd! A working document that was treated as confidential! As if the competition were going to steal it and take all the business away! The company was already doing such a good job driving their customers away to the competition thanks to their late deliveries that giving away the schedule would have had no real effect.

The shop floor scheduling system was working in a vacuum. They had a system, all right, but the different departments did not see the overall master schedule and the shop floor schedule (confidential, you remember). And, because of the lack of communication, the fabrication department for weldments, the machine shop, and the assembly floor did not communicate, so there were schedule priorities only for each individual department. They were doing the best they could, given their lack of information. Each department was producing, engineering, fabricating, machining, and assembling to its own schedule. While each department knew what the due date was, they did not know the capacity requirements of the other departments and could only schedule to maximize their own department.

How did we fix this situation? The company was ready to junk the MRP system and get a new one, because they had proof that the system did not work (sound familiar?) — they had firsthand experience watching a system not working. The solution was as basic as could be. All they needed to do was make the master schedule available to everyone in the company which allowed them to master schedule the entire job. That took care of the planning portion of the operation (time frame: one week). Then, when they saw what the schedule was, they could properly sequence each of the four departments that had an effect on delivery time, as they were the units responsible for the consumption of the important part of capacity, time.

After scheduling to capacity, they sequenced the jobs through effective dispatching to the shop floor and managed the sequence daily on the floor.

Each morning, a scheduling meeting reviewed the previous day's progress and discussed schedule changes and that day's objective. It took about two months to get this procedure running smoothly, to reschedule late orders, and to get the operation back on track. And the company got rid of the dreaded "hot list" (more on hot lists later in this book). All they did was something as simple as managing the timing and sequencing of the jobs as they went through the floor. The main tool they used was bottleneck scheduling, as each month a different machine would become capacity critical because of the mix of jobs in the shop. The result of their efforts was a simple solution to a problem that had been plaguing the company for years.

Shortly afterward, the president of the operation called to say thanks but also to report that the fighting over the scheduling of the shop floor still continued. All that work, and the fighting still raged on! He was halfway kidding, though. The fighting was, indeed, still going on, but now it was being done in the master scheduling meeting, where they were fighting for time slots in the schedule to meet the promise dates of the customer. They were allocating the most precious commodity in the company — time. And it is at the master schedule meeting where decisions on time allocation should be made. The rest is schedule attainment.

Why Teach Basics When the Answers Are Out There?

Right now, somewhere out there in industry is a manager or a management team ready to sign up for the latest technology (MRP, MRPII, ERP, Supply Chain Management, Real Time Shop Floor Control — you name it; see Figure 1). With their signature on the contract, they are about to commit the firm and its resources to a major undertaking. In many cases, the odds are going to be against their succeeding. Let's offer you this one fact: Current analysis states that somewhere between 90 and 95% of all new manufacturing software installations fail. This figure includes not only the initial startups but also implementations that represent second and third tries. Stated in a different way, almost all the startups of new manufacturing systems fail when they are first turned on.

It is too easy to blame the software ("we bought the wrong package") or the software vendors ("they lied to us, but then what did you expect?"). Yet, for most companies, this is not the case. All that the company is doing is shifting the blame from the guilty to the innocent (or not as guilty). The real root cause in these situations is that they did not do the basics "right."

Ask yourself this question: "Why is it that the software failed in my operation?" It worked all right in your competitor's plant right down the street. What is it that your competitor did that you didn't that made the system work for him? The answer is simple. He got the basics right. He did his homework and put all his ducks in a row. All the tools needed to run the new system were ready to go and were in place. We meet a lot of people who say that the software that

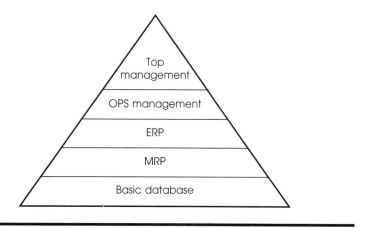

FIGURE 1. Operations

they bought a few years ago is not any good — the software implementers did not know what they were doing and the stuff is junk anyway, so the software has been sitting on the shelf. To solve the problem now they are going to buy a "good" system that they know will work. Guess what? All they are going to do is pick another "bad" software package that will also fail. That is the "why" of this book.

The solutions to your problems do not lie in software; they can be found in "basicware" — knowing and consistently and effectively applying the basics. So often, all the pieces are there, just as they were for the company discussed above, but something very simple and basic is wrong. We have talked to a lot of people who say that the computer system they have is not working and they are going to buy a new one. Even though the guys across the street have the very same system doing basically the same thing, these people never think to ask themselves, "What are they doing that I'm not that is making the system work for them?"

The ones who are succeeding do the basics well. To these people, the basics are just that — basic. The basics are unexciting, the basics are known, the basics don't get you written up in *Fortune* or *Business Week*. All that the basics do is help you make money and keep your customers (and your employees) happy. We find that it is the really basic solutions to complex issues facing your company that can get you turned around. Read on if you want to find out how to be world class.

Basics Are Hard, But Basics Work!

Today we have another problem that compounds the issue. As recently as two years ago, if we bought software that failed to work for us, we could fall back on the old system to keep running. Not today. There is a high probability that the old software will not run again. In light of Y2K, it is most likely that our old system will shut down, so we have to make the new software run. We do not have a choice. If we fail in our implementation, we are back to square one, running our operation in a manual mode. But, if we can't get the new system to work and the old system is obsolete, whether you want it to be or not, we are stuck. We are now in a fight for our corporate life. We are facing the enemy, and he is us, a reality which is very nearly a law of manufacturing and is frequently referred to as the Pogo Law, named after the old cartoon character: "We have met the enemy and he is us." We are focused on the desired outcomes and have the will to win, but we are failing to consider the individual battles. Wars are won one battle at a time, and if we focus on the war, rather than the battles, then basics tend to become the debris of battle. The stuff is all around us. And the pile of debris is growing as we fight our way through the implementation.

When we excel at the little things, everything gets easy. Why? Because it is the basics that we do, the little things, that make us a success … the little things such as doing a really good job on creating the most accurate bill of materials that we can. Something as simple and basic as getting the bill of materials right has a tremendous impact on your operation. Basics bridge the gap between theory and practice. The theory shapes the operational side of the equation. It is the theory that defines the operational parameters of the system that we need to install to acquire the manufacturing technology inputs that are needed in our corporation.

It was Santa Anna who said, "Those who are ignorant of history are doomed to repeat it." This observation makes sense here, too. If we do not look at getting our database right and understand the essentials of capacity and scheduling, we will again fail to implement. As we said earlier, there is a high failure rate for the initial implementation of new system software. Are we going to ignore the lesson learned here and not go back to review what we did wrong the last time? Are we going to buy another new software package, as the last one must not have been any good because it failed? If we do not look at the reason why the first system implementation failed, then, as Santa Anna said, we are doomed to repeat the mistake. The lesson is a simple one. We must do a much better job of managing our databases and the integrity of the inputs that we use in our operating systems.

Today, in the information age, we need to do a much better job of managing our data integrity. What we thought was a good database last year is in no way acceptable today. The speed at which the new systems operate and the levels of integration that they require literally force us into maintaining nearly flawless databases and operating parameters to make these new millennium software systems work. Coca-Cola has developed what they call a smart vending machine. They have realized there are tremendous savings to be gained from managing the operation at the can level instead of at the batch level, or at the daily or monthly level of production. They want a smart vending machine that can trigger the replenishment cycle to manage the business at the can level; there are significant gains to be had by managing the business at that low level, but we must have the basics right in order for our operations to function at that basic a level.

This is why we wrote the book. To give you, the manufacturing executive, the tools and the perspectives necessary to think at the can level. To give you the tools to make your operation a success and the basic skill levels to manage the supply chain with the technology available to you. You cannot simply buy a new system and assume that it will work. Success comes cleverly disguised as hard work, and that hard work is also known as the basics.

Before finishing with this introduction, let's be clear about one thing — knowing and doing the basics well are not easy. They require work, understanding, and commitment. Focusing on the basics — on the things we must do well every day — is not a quick fix. It requires knowledge. Giving you that knowledge is the purpose of this book, as it has been our experience that for every minute spent on understanding and applying the basics and for every dollar spent on the basics, there is usually a tenfold payback. If you are ready to learn the basics and to benefit from the gains offered by the basics, then this book will help you learn.

Remember, a lot of problems can be solved by going back to the basics, and a lot of problems can be avoided if we rely on the basics in the first place. That is the promise of basics — doing what we must do with excellence. And that is why you are reading this book.

Mapping Manufacturing Excellence: The Case of the MRPII System That Wasn't

Overview

One of the most common questions that we receive deals with inquiries from various readers regarding recommendations on various new developments that the readers have heard about. Inevitably, the readers have attended a seminar where they have been introduced to one or more new technology advances. These advances include such developments as Just-in-Time, e-commerce, interactive scheduling, MRP, MRPII, or ERP. Before we proceed any further, it is important to note that nearly all of these new technological developments can and will work and generate significant results, provided that they are used in the proper settings to address appropriate problems and that the correct set of basics is already in place. Most of the people who call us are immediately disappointed when we ask them about the various features of their systems and the extent to which their production systems have in place most of the basics demanded by these new developments. Somehow, these managers have become convinced that they need not focus on the basics and that by taking advantage of this new development all of their problems will be solved. The reality for most of these managers is that, in most cases, these new advances will do little more than consume time, manpower, and resources with little to show in return.

What managers must realize is that manufacturing excellence, no matter how it is defined (cost, quality, lead time, flexibility, or innovation), is not a function of one new technology or development; rather, it is the outcome of a large number of small things — each carried out well and each well integrated with the other components. Sometimes, management, by realizing this essential fact of manufacturing, can avoid a potentially large and costly mistake — which is what one firm that we know of did.

The Case of the MRPII Implementation That Wasn't

Some 10 years ago, a firm located in the midwest contacted one of the authors with a simple but urgent request. The top management had attended a seminar on Manufacturing Resources Planning (MRPII) that had been put on by a major software vendor. After sitting through some two days of presentation, the management team was convinced that this was the solution to various problems that had long plagued the manufacturing system. They immediately invited the vendor to send a team to evaluate operations at their location and to assess the suitability of the vendor's package for use in their manufacturing system. Within the month, a four-member team (two consultants and two software experts) arrived and spent over three days assessing operations within the plant. The conclusion of this team was that, yes, performance could be better at the plant and, yes, their MRPII product was just what was needed. Within 30 days after their visit, the company received the vendor's formal proposal. The only thing that bothered them was the cost — it was higher than what they had expected. However, if it worked, then there was no need for worry because the implementation would pay for itself within 18 months. This proposal was presented to the Board of Directors, and it was conditionally implemented. The condition that was added by several of the board members was that the proposal and the operations must be evaluated by an outside source. If this source agreed that MRPII was indeed the answer, then the contract could be approved.

That is why representatives of this firm decided to approach several members of the Business School with a request that they come in to evaluate their operations. A date was arranged for the Business School team to visit the firm and listen to presentations from both management representatives and the vendors as to why MRPII (that specific vendor's version of MRPII) was most appropriate for the plant. After listening to the presentations, the team went out on the floor to evaluate operations. What they saw there raised doubts about the firm and its manufacturing system.

The first place that the team visited was the stockroom. They asked about inventory accuracy, only to be told that no one really measured it. No one was too worried about inventory accuracy, as they had never stocked-out of any order as far back as they could remember. At that point, one of the management team invited the team members to visit the new warehouse that was being built near the plant (the purpose of which was to store the excess inventory that the plant was experiencing). The team also noticed that there was no real control over the inventory — anyone who needed inventory could walk in and take what they needed and then walk out. During the process of assessing the stockroom, one of the team members noticed a pallet of items partially hidden in one corner. The pallet had a bright red tag attached to it, informing everyone that this pallet was urgent — the date on the tag was June. That wouldn't have been too bad except for the fact that it was currently the middle of December.

During their assessment of operations, the team noted the following:

- Capacity was never considered when doing planning (many of the planners could not even hazard a guess at the capacity of their system).
- All production scheduling was done manually on a large, vinyl-coated metal board.
- Many of the items in the system had no real formal bills of materials and those that did had varying levels of accuracy.
- Routings for many parts were missing (people depended on the memory of certain key individuals), and the routings for many of the remaining parts had not been updated in many years.
- No one on the floor really believed in the proposed system or even understood it.
- Expediting was the norm, rather than the exception. This was a system in which the "squeaking wheel" really did get all of the attention.

After spending time in reviewing the system and its operation, the Business School team reconvened with the management representatives. Their recommendation was that, while MRPII was appropriate, the firm was not yet ready for MRPII. Most of the basics demanded by any MRPII system (and most successful manufacturing systems) were not in place. Before MRPII could be implemented, these basics had to be put in place.

Based on this recommendation, the company decided to delay the implementation of the MRPII system and to focus on making sure that the basics of manufacturing were done right. After working intensively on the basics for some 12 months, the firm moved ahead with its decision to implement the MRPII system. The system was implemented without problems in less than 12 months, and the firm quickly became recognized as a leading class-A MRPII user.

Mapping the Basics of Manufacturing Excellence

There is no one single key to manufacturing excellence. Rather, manufacturing excellence is composed of many interwoven elements or strands. These various strands are summarized in Figure 2. What this map shows is that there are numerous activities and tasks that must be completed before excellence can be achieved. It also shows that different groups in the firm are responsible for different activities. For example, business planning and sales and operations planning are the responsibilities of top management. It is their job to make sure that the plans pursued by both marketing and manufacturing are consistent not only with each other, but also with the corporate strategic plan. Further-more, they are responsible for making sure that there are sufficient resources (i.e., capacity) present so that these plans can be realistically executed. Before these plans can go to operations planning, they must be evaluated for adequacy of resources.

The operations planning stage encompasses master scheduling (one of the most critical manufacturing plans found within any firm), material planning (to ensure that sufficient materials are available), and capacity planning. Op-erations planning works within the constraints established at the strategic planning level, and its goal is to generate feasible, valid schedules, which are then carried out within the execution system. Operations planning must make sure that these schedules are consistent with the strategic plans, as well as with capacity, inventory, and material lead times. This planning is far more detailed compared to that found in the strategic planning level.

At the third level, we have execution. Here, we are talking about imple-menting the plans generated and evaluated within the preceding levels. These plans are implemented either within the factory (our own manufacturing facilities) or within the external plant (capacity found within our supplier base that we can draw upon when and if we need it). This level is the most con-strained, as it must work within the timing and resource constraints established at the preceding stages. Further, it is the most detailed of all three stages. It has to be detailed, as we are now dealing with issues of who will do what in what order and when and where.

Completing this process or structured set of activities is performance mea-surement. Measurement is critical, as it helps communicate what is important, how well we are doing, and the opportunities (and relative priorities of oppor-tunities) for improvement.

Unifying this entire process are feedback loops. Within the feedback loops, the major resource being moved is that of information. This is an important linkage, as it helps the entire process and all of its players be aware of the progress of work and of changes taking place within the system. Some of these changes are critical because they affect the feasibility of the plans. For example, if a vendor experiences a major problem, such as a labor disruption (strike or work slowdown) or a loss of capacity (e.g., due to a plant fire), then this loss of

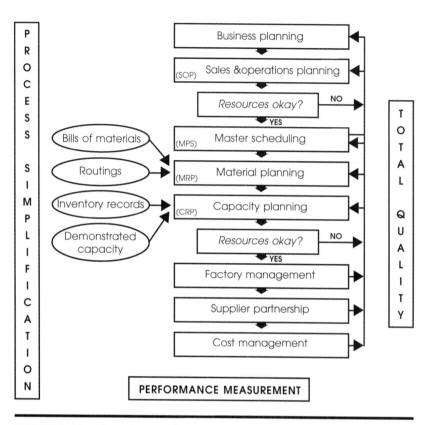

FIGURE 2. Manufacturing Excellence Map

capacity may mean that plans previously deemed to be feasible are no longer feasible. Replanning must be done. This replanning can be done at the master scheduling level (operations planning), which is usually the starting point for such replanning. It is important to remember that as we proceed up the map we are dealing with a greater number of groups, and we are dealing with issues that require extensive renegotiation if they are to be changed. If the feasibility of the schedule cannot be established at the operations planning level, then it must be reexamined at the strategic planning level, where changes can impact the strategic stance of the firm.

In reviewing this map, we see that there are many things that we must do well. We cannot hope to focus on one area and ignore the rest. Developing wonderful master schedules that are not feasible from either a material or capacity perspective does not help the firm achieve its objectives. There are four areas on the map that deserve special attention — these are bills of materials, inventory records, routings, and demonstrated capacity. They will be covered in the next section.

Mapping Manufacturing Excellence: The Four Operations Pillars of Excellence

Overview

Manufacturing excellence, as we have previously shown, is built around a map of excellence. This map identifies the various activities that we must carry out and the sequence in which we must do so. It also shows the various linkages that tie together these various activities into one integrated whole. However, this map, while identifying the various functions, also sets out the operations tools or pillars that are critical to getting the job of effective and successful planning and execution done. These are operations tools, as they are the ones that the production and inventory control manager most often draws upon, controls, and manages.

The Four Operations Pillars of Excellence

These four pillars, identified on the left side of Figure 2, are the tools needed to get the manufacturing tasks done correctly. These four tools must be accurate, and they must be complete and current. Failure to have these pillars in place will prevent manufacturing from doing its job correctly.

Following are the four basic tools:

1. The *bill of materials* (BoM) is that portion of the manufacturing database that tells everyone involved and the systems in place what parts are needed to build items. We can do no planning, scheduling, ordering, or order promising unless we know the status of the required parts for the job.

2. The *route sheet* (or router or process sheet) is that portion of the manufacturing database that specifies the sequence of assembly and how long it will take to get a job done. We cannot promise our customer a date for receiving their order or know how to put the product together without accurate data from the route sheet.

3. *Inventory record accuracy* answers the first question you ask yourself when there is a requirement for a product: "How many do I have?" If you do not know how many parts you have in inventory or how many finished goods are available to ship, then you cannot determine a delivery date. And we are not talking about percentage accuracy of your inventory. If you have just one record that does not accurately reflect the amount of materials or finished good that you have, the credibility of the entire inventory is in question.

4. *Demonstrated capacity* is what you can actually produce and what has been demonstrated over time as being achievable. This is not the place to lie to yourself regarding your capabilities. You must be aware of the

available time that is at your disposal to complete orders. You also need to know how much of that time is committed to other orders. Knowing these two facts, you can then determine when the order will be completed, subject to parts availability.

These are the four data pillars that are necessary to achieve manufacturing excellence. They are as basic as they get. And we have to do all of them right. And we must do all of them completely in order to have a viable database and to achieve effective and successful planning and execution. We cannot schedule if we have only a percentage of the parts listed on the BoM. We cannot schedule unless we have a complete understanding of the capacity of all the equipment available to us. We cannot schedule unless we know how much time it is going to take to complete the order, and we cannot tell how much we have in inventory and where it is unless we have all items involved in our operation located accurately.

In regard to the basics, it is an all-or-none situation. Here is something to think about. Considering the ability of today's computer operating systems to process data, and the extent to which they can handle and manipulate these data, any discrepancy in the accuracy, completeness, or integrity of data in our database could cause the entire computer-based manufacturing system to crash.

Lead Time: One of the Basic Concepts of Manufacturing

Overview

Almost everything that we do in production and inventory control involves lead time in one way or another, as every order that we process has a due date associated with it. We are under constant pressure to reduce lead time and are told that we must have more predictable lead times. And, of course, there is the popular notion that "time is money." Yet, in spite of its criticality to production and inventory control, lead time is not a well-understood concept. For many, it is a source of confusion, yet this should not be the case. Lead time, while potentially complex, is relatively straightforward to understand, once we take the time and effort to think about it. Given its importance to manufacturing, we will now examine this concept.

Defining the Concept of Lead Time

A logical starting point is to define what we mean by lead time. Lead time can be defined as the elapsed time interval between the start and end of an activity or series of activities. Lead time can be examined in one of two ways. We can look at the lead time of a specific event or order — for example, we can talk about how long it took to complete an order or a specific operation or process. Alternatively, we can look at activities or orders over a given period of time (such as a quarter or a year). When we take this approach, we have to talk in terms of distributions. With a distribution, we must worry about issues such as the mean (the average lead time for an order or a process) and the spread (the difference between the worst and best cases in terms of lead time), which is an indicator of predictability — the greater the spread, the less predictable the lead time.

Types of Lead Times

Having defined lead time, we must differentiate among some of the terms frequently associated with lead time, such as speed, reliability, and predictability. While often used interchangeably, these three terms have very different meanings. *Speed*, for example, refers to the elapsed time that it takes to complete an operation or to fill an order. Typically, we refer to the average lead time when we are talking about speed — for example, it takes 15 minutes to compete a particular step; implied here is that this is the average lead time. *Reliability* refers to the extent to which we can meet requested due dates. Anyone who flies is familiar with this concept, because it is the one that reflects the ability of an airline to provide on-time departures and arrivals. Finally, *predictability* refers to the amount of variance or spread in the lead-time distribution. In general, as previously noted, the greater the difference between the shortest observed lead time and the longest observed lead time, the more unpredictable the lead time. Typically, we evaluate predictability using such statistical terms as variance or standard deviation. For those of you who are statistically challenged, one way of understanding the standard deviation is to compare it with the mean. If the mean tells us what we can expect the lead time to be, on average, then the standard deviation tells us how far off the mean, on average, we will be if we are wrong. When we work on reducing the spread or the standard deviation of a lead-time distribution, then what we are doing is essentially making the lead times more predictable.

In addition to these dimensions, we should also differentiate between the planned lead time and the actual lead time. Systems such as Material Requirements Planning (MRP) and Manufacturing Resources Planning (MRPII) require and use planned lead times. Typically, a planned lead time is a value that we provide to the system to be used for planning purposes. In most cases, the planned lead time is expressed as a point value. Even though we know that it is taken from a distribution, most computer systems cannot deal with a distribution. While the lead time for a specific activity, such as building one unit of a component, can take anywhere from two to four weeks, most computer systems want one number (e.g., three weeks). The planned lead time can be viewed as a forecast. When we provide a value for a planned lead time, we are telling the system how long we expect it to take to make one unit. Because it is a forecast, we must recognize that we have to deal with the possibilities of being wrong (of having the forecast disagree with the actual). If these errors occur, it is important to note that we need not pick the mean as our planned lead time. In some cases, we may select a value greater than the mean. When we do so, we are implicitly planning for inventory. Why? Because, if it takes three weeks, on average, to make one unit of output and we plan for four weeks, then an order being completed in three weeks will sit in inventory for one week before it is used.

It is important to remember that there is a tendency to set planned lead times at a pessimistic rather than an average or optimistic value. The reason is that an order being late (for whatever reason) can potentially delay other orders that are waiting on that part before production can proceed. In these situations, it is often seen as being better to arrive "early" and sit in inventory than to arrive late.

While the planned lead time is a forecast, the actual lead time is the specific actual elapsed time that we observe. It reflects what actually took place. It is not a forecast. This brings up an interesting but important issue — that of dealing with the gap between the planned and actual lead times. We know that there is no problem if the actual lead time agrees with the planned lead time. Yet, if the actual lead time exceeds the planned lead time, we are faced with a late order. On the other hand, if the actual lead time is less than the planned lead time, we have temporary excess inventory. When these gaps take place, we can adjust the planned lead time (either upwards or downwards) to close the gap. However, in most cases, this is the wrong thing to do. If we adjust the planned lead times in response to changes in the actual or observed lead times, we see "dial twiddling" in action. This action disrupts the operation of any planning system and is to be avoided at all costs. Rather, we must try to start out with reasonable lead times and then manage the actual lead times so that they are in agreement with the planned values.

To achieve this balance between the planned and actual lead times, we must try to study the factors and processes that affect the actual lead times and we must get the factors under control. If the excess variability that we see in the actual lead times is due to factors such as uncontrolled releases of work to the shop floor, then we have to study the process by which work is released to the shop floor for implementation, and we have to eliminate the problems present in this process. If the problems are due to large lot sizes, then we have to reduce the lot sizes by reducing the setup times (a major contributor to large order quantities). Variability is a symptom of a problem in one or more processes. We must address the underlying causes, not the symptoms.

Major Lessons Learned

In this section, we have focused our attention on the critical concept of lead times, and we have defined what is meant by the term. We have shown that lead times are best described by distributions. We have examined the differences among terms associated with lead time, such as speed, reliability, predictability, and planned and actual lead times. The only dimension of lead time that we have not discussed is its components, which we will examine in the next section.

Understanding the Components of Manufacturing Lead Time

Overview

Manufacturing lead times are very important to all production and inventory control (PIC) systems. They influence nearly every aspect of the planning and execution activities that take place within PIC systems. Manufacturing lead times (both planned and actual) are used when planning capacity, material, and priority needs. These lead times influence the amount of work-in-process inventory present. There is a direct and positive correlation between the level of work-in-process and lead times. As a general rule, a doubling of manufacturing lead times will cause a doubling in the level of work-in-process inventory. In addition, lead times affect safety stock levels and the reorder points (if used). Finally, manufacturing lead times can and do influence the firm's strategic stance, for as manufacturing lead times increase, the firm's flexibility (and attractiveness to its customers) decreases. It is a basic rule of manufacturing that, for the firm to be successful, the PIC system must manage and control manufacturing lead times. However, before we can begin the process of managing and controlling this lead time, we must first understand what we mean by the concept of manufacturing lead times.

Defining the Concept of Manufacturing Lead Times

Manufacturing lead times can be defined as the time that elapses between the release of the order to the shop floor and the completion of all operations on the order and its receipt into stock. Manufacturing lead time can be divided into two basic categories:

1. *Operation time*: The time during which the order is processed (i.e., being worked on). Included in this time component are such activities as setup (discussed elsewhere in this book) and production time. It is important to know that, of all the elements of manufacturing lead time, the only one that inherently generates value to the customer is that of production time. It is only during production time that we are changing the product into something that more closely parallels what the customer wants (and is willing to pay for). The other activities, while potentially important, do not have as great a capacity for generating

value. That is one reason why some firms and managers measure the total number of sets in a process (or in manufacturing lead time) and the percentage accounted for by processing time. As a general rule, we can assume that the greater the percentage of total lead time accounted for by processing time, the more effective and efficient the underlying process.

2. *Interoperation time:* The total time that the order is involved in non-production activities. Stated in another way, this is the total time that the order is not involved in operation time.

In most systems, interoperation time often accounts for the largest percentage of manufacturing lead time. In part, this is due to the fact that interoperation time consists of numerous elements, each of which has to be identified and understood:

- *Queue time:* The time that the order spends waiting at a machine or a work center before being processed.
- *Preparation time:* The time required to complete work that must be done before the operation can begin. Included in this component are such activities as the time spent in cleaning, heating, or marking out the part. We would even include the time spent in pulling out blueprints or reviewing drawings or customer instructions.
- *Postoperation time:* The total time that takes place at the end of an operation and creates no load on the work center (i.e., it consumes time but not machine capacity). Examples of activities that make up this time component include cooling, cleaning, deburring, wrapping, and local inspection.
- *Wait time:* The total time the order spends waiting for transportation to the next work center or work area.
- *Transportation time:* The total time that the order spends moving between work centers.

Of these various components, queue time is typically the largest and most important element. In many environments, queue time accounts for between 75 and 95% of the total manufacturing lead time for a typical order. The general composition of manufacturing lead time is shown in Figure 3.

Factors Affecting Manufacturing Lead Time

In reviewing the various components that make up lead time, it is important to note that they are influenced by different factors. A component such as production time is significantly influenced by such factors as the size of the order (i.e., order quantity), the time required per unit, the condition of the equipment (which affects the probability of a breakdown), and the expertise of the person

Queue Time	Preparation Time	Setup Time	Production Time	Post-operation Time	Wait Time	Transpor-tation Time
Interoperation Time		Operation Time		Interoperation Time		

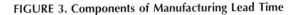

Manufacturing Lead Time

FIGURE 3. Components of Manufacturing Lead Time

doing the processing. Operation time as a total component of the overall manufacturing lead time is affected by such factors as the total number of operations to be completed and the order quantity. In contrast, interoperation time depends on factors that are often not order related. For example, queue time is dependent on the number of jobs ahead of the current order at the specific work center, the time required to complete these other orders, and the dispatching/scheduling rules present. Wait time is dependent on the amount of capacity available for moving orders between operations, the number of other orders waiting, and distance between work stations (to name a few factors). In general, interoperation time is longer in duration than operation time. More important, it is also the major source of lead time variance. These variances indicate problems with such areas as loading, scheduling, and work practices. To reduce variance in these components, we must examine the entire manufacturing process and the systems that influence the operation of this process.

Before leaving this section, we should emphasize a point previously made. Interoperation time presents a twofold problem. First, it is large in size. Second, it often embodies activities that increase cost more than they increase value. As a general rule of thumb, we want to reduce the amount of time the order spends in interoperation time. This is time that creates costs without providing any real offsetting value.

Importance of Monitoring the Elements of Manufacturing Lead Time

Having identified the major components of the manufacturing lead times, the next issue to address is that of how to use this information. One way that this model of manufacturing lead time can be used is to help structure data collection. That is, it is useful and insightful for us to collect information about: (1)

how long it takes us to process an order from start to finish, (2) where the order spends its time (i.e., how much of the total time is spent in queue time, processing time, or wait time), and (3) the amount of variance experienced overall and by component. Why measure manufacturing lead time, in general, and its various components, specifically? The answer is simple — we cannot control what we do not measure. Without measuring manufacturing lead time and its various components, we do not know what is happening to manufacturing lead time or what factors are influencing its performance. Furthermore, by measuring these components, we identify areas to focus on and opportunities for improvement.

By reducing manufacturing lead times (either the average or the variance), we generate significant benefits for both the production and inventory control system and the firm as a whole. With shorter lead times, inventories fall; with shorter lead times, system responsiveness improves; with shorter lead times, costs fall and quality improves. It is for these and other reasons that manufacturing lead time and its control are fundamentals of production and inventory control.

Major Lessons Learned

In this section, we have focused our attention on manufacturing lead time. We have defined the concept and have shown why manufacturing lead times are so critical to the firm. Further, we have broken down manufacturing lead time into its various components. We have shown how we can regard these components of manufacturing lead time in terms of their impact on costs and value. Finally, we have explained the need for an ongoing measurement and monitoring of manufacturing lead time and its components. As production and inventory control managers, we are resource managers; we are also lead time managers. One of our products is that of manufacturing lead time.

Manufacturing Is Really Just a Balancing Act

Overview

In order to understand the different elements in our manufacturing operation, we must begin by understanding the interrelations among the functions that make our operation run. The best way to visualize this is to imagine our operation as being a balancing act between the various components of our operation, our systems, and our operational capabilities. Upset this balance, and problems begin to arise in our operation. Keep the elements in balance, and all should run fine. Understanding how these functions work will help you to understand solutions to the problems that we face in our operations and how the solutions affect the outcomes.

The Balance

Take a look at Figure 4. We have a balance beam that represents our operation. It is a simple balance beam, not that different from a balance scale. On the left side we have our systems. These are the tools that we use to run our manufacturing. These are our sales plans, our computer system, our suppliers' capabilities, our forecast system, our customers' requirements, and transportation issues. All the items in the systems box on the beam are the issues or constraints that we must deal with to run our operations.

On the other side of the balance beam is the box representing operations in which our manufacturing capabilities reside. This box contains our production capabilities, our available capacity, our throughput processing capabilities, our manufacturing lead time, our capacity constraints, our inventory record accuracy, the accuracy and completeness of our bills of materials, and our route sheets.

Now, looking at our manufacturing system, we will begin with the premise that when each side is equal we are in balance and all is fine … sort of like the teeter-totter we played on at the park when we were kids. When your weight was the same as the person on the other end of the beam, the beam remained stationary in a horizontal position and you were in balance. If your friend was larger than you, then that end of the beam went down while you went up and were trapped high in the air. The beam was out of balance and no longer functional. If you had a really big friend, there was no way for you to rock the beam up and down, as the beam was way out of balance. To solve this problem, you could have had another friend of yours climb on the beam at your end so your combined weight could bring the teeter-totter back into balance.

Applying this analogy to our manufacturing operations, if our systems and our operational capabilities are in sync with each other, the beam is in balance, our operations are in sync with each other, and all would be running fine. But, if our system capabilities are not in balance with our operational capabilities, then the beam tips. When our manufacturing system is not in balance with

FIGURE 4. The Balance Beam of Manufacuring — the Basic Components

operations, we can easily see what the effect is. We begin to experience longer lead times, stock-outs, missed shipments, or, worse, lost customers. The system is out of balance and we are experiencing problems.

To get ourselves back in balance, we can again turn to the playground example. When we had our friend climb on the beam with us and balance the weight, we were then back in balance and all was working well. Well, in the manufacturing arena we also have a friend who we can add to the beam to add weight to the light side and get us back into balance. That friend is called inventory. Inventory is the weight that we add to our operation to bring us back into balance so that everything is back in sync again. It can be placed on either side of the beam as necessary to regain balance. It can be used to add weight to weak systems and weak operational capabilities. In essence, the inventory box can be moved to wherever it is needed, anywhere on the beam. If placement of the box cannot add enough leverage to balance the beam, then we can add a bigger box for more inventory. This now begins to explain the quantity of inventory we have in our operations and why we even have inventory. Inventory is a universal balance equalizer. Inventory supports the areas of our operations that are weak and is essential for keeping us in balance.

Look at Figure 5 to see how we have added weight to the beam in the form of inventory. Let's assume that our system requires us to produce and ship an order to our customer in five working days. We can do it in five days, so everything is fine, and the system is in balance. But, if the customer wants the order in three days and we still need five days to deliver, then we are out of balance and cannot make the delivery. If we cannot produce and ship in the time required, then we have only two options. The first is to turn down the order. Obviously, this is not acceptable. The second option is that we add inventory to the system so that we are now in a position to meet the customer's shipping demand of three days. Because we are unable to meet the customer's demand for three-day delivery with our present manufacturing and inventory policy, we must increase our inventory as a short-term solution to the problem. We ship from that inventory, and the need to stay in balance begins to

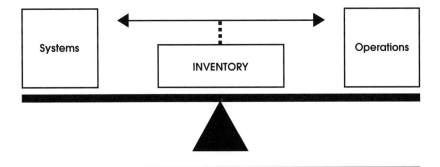

FIGURE 5. The Balance Beam — Positioning Inventory

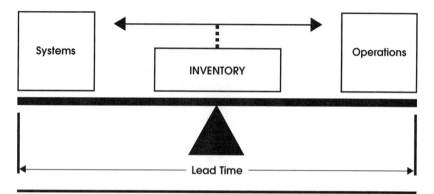

FIGURE 6. The Balance Beam — Understanding the Role of Lead Time

determine the amount of inventory that is necessary to keep on hand to meet the needs and requirements of the customer. The size of the gap between what our customer wants and our ability to deliver dictates the amount of inventory that we must keep on hand. The long-term solution is to do something about the size of the systems box or the operations box to enhance the robustness of the weak link in the delivery chain to meet the three-day delivery window, but that is a long-term fix and could be quite costly.

If you remember your physics, length times weight equals mass, or, in plain English, multiplying the weight of the box by the distance from the box to the balance point determines how much weight is acting on the beam. This tells us how much weight is necessary and where to place the weight on the other side of the beam to keep it in balance. From this, we can see that a weak aspect of our operation can be brought back into balance just by moving the weak box farther away from the balance point and lengthening the beam until we are in balance again. While this strategy would work in theory, in reality we have a name for the length of the beam — lead time. If we move the weak link farther away from the balance point on the beam by increasing lead times, we do, in fact, bring the beam back into balance. But we do so at a cost, and that cost is the amount of lead time necessary to deliver.

If business requirements have weakened a link in the system, we could bring the beam back into balance by adding length to the beam and moving one of the boxes farther out from the balance point until the system is balanced again, but our lead time has now increased. Now that lead time has been added to the balance beam, we have a balance beam that looks like Figure 6.

We have now identified all of the components of the beam that represents our operational capabilities. Now we can clearly see what happens to the system when we try to meet the customer's three-day delivery requirement with a five-day delivery operation. We can approach the delivery requirement in two ways over the short term — lengthen the beam and keep the five-day window, or add inventory and make the three-day window. In each case, we now notice that if we change one of the parameters of the balance beam we will need to change another parameter to keep the beam in balance and meet our objectives. What we see now is that there is a cause-and-effect relationship to consider when trying to keep the beam in balance.

Now, because there is a cause-and-effect relationship among the factors on the beam, we realize that changing one of the elements of the beam means that another element on the beam must also change to keep the beam in balance and our operation in sync. When we were required to shorten the lead time to the customer from five days to three, we found that we could take one of two actions. The first would be to add inventory to the system, which would keep the operation in balance and meet the customer's requirement by allowing us to ship from that inventory. Here is the cause-and-effect relationship for this alternative: If we add inventory to shorten lead time, then we attain equilibrium. On the other hand, we could leave the inventory alone and maintain a five-day lead time because that is where the system is at equilibrium. If we eliminated more or all of the inventory in our system, then we would end up with a balance requiring an even longer lead time.

These are short-term considerations. In the long term, we need to assess what our needs are and the delivery window necessary to meet our customers' requirements, then make the changes necessary to meet the new demands and keep our system in balance. Inventory is the tool we use to keep the system in balance. For example, if our goal is to reduce inventory and we install a new order-processing system that makes the system box more robust, we could then reduce inventory. But, if our sales force starts to promise shorter delivery lead times to their customers based on the efficiency of the new system, we can now see, by looking at the balance of the operation, that our goal of inventory reduction is in jeopardy.

Major Lessons Learned

We have a project for you. Take out a piece of paper and a pencil and develop your own balance beam for your company. Write down all those items that would fall in your systems box. Then write down all those items that would fall in your operations box. Calculate the lead time you need to deliver to your customers. Now you have the balance beam set up. But, we still have not included inventory on your balance beam, because inventory is the only function you have that you can easily manipulate by either increasing or decreasing it and moving it around on the beam. That is why we saved inventory as the last piece for you to work on. Inventory is reactive to the balance of the beam, not proactive, as inventory reacts to the need of the system to stay in balance.

This is why we have inventory. It is our reaction to the necessity of maintaining balance in our operations. So, the next time you begin a quest to eliminate inventory, understand the effect that an inventory reduction program will have on the balance of the operation.

2 Understanding Management: The Basics of Problem-Solving

Why Problem-Solving?

Managers do more than simply manage. They are intensely involved in the process of directing resources and organizing activities in achievement of corporate or organizational objectives. Ultimately, what this means is that managers are problem-solvers and decision-makers. They must be able to bring order to chaotic situations. Furthermore, they must be able to examine the situations facing them and distinguish between the symptoms and the underlying root causes. They must avoid the tendency of attacking symptoms and instead focus their attention on the problems.

The problem with problem-solving and decision-making is that, for most managers, these are activities and processes that have been learned only after a great deal of experience and postgraduate work in the school of "hard knocks." This approach allows the new manager to learn the necessary skills of problem-solving only after numerous periods of trial and error. This process is time-consuming, costly, and painful (for the new manager, the firm, and those working with that person). What this approach overlooks is that there are certain skills that can be learned by reading, rather than by trial and error.

In this chapter, we provide you, the reader, with an overview of some of the critical skills, concepts, and processes of problem-solving. It is hoped that, with this knowledge in hand, the new manager can spend less time making errors and more time helping the firm achieve its various goals and objectives.

Understanding the Concept of Manufacturing's Most Fragile Asset: Information

Overview

No discussion of the basics of operations would be complete without a discussion of data and information. Beginning with the introduction of Material Requirements Planning (MRP) in the mid-1970s and the combined effect of falling computer prices and increasing computer power, we have seen a movement from management relying on inventory toward management driven by information. We now rely on information and data to schedule orders, meet commitments, respond to customer requests, assess performance, and identify potential manufacturing needs. The reductions in inventory that we are now seeing are due, in part, to the increasingly important role played by information.

As we move from focusing on the internal factory (our own facilities) to focusing on the supply chain and the management and coordination of both the internal and external (suppliers' facilities) factories, we can expect information to play an ever greater role. Developments such as MRP, Manufacturing Resources Planning (MRPII), and Enterprise Resource Planning (ERP) are fundamentally information driven. As we proceed down this path of greater reliance on information, it is important that we develop a good understanding of the basics of information. Knowing and understanding information are as critical to manufacturing excellence as our knowledge of capacity, bills of materials, routings, and inventory records.

In studying information, we must focus on four areas:

1. Defining information
2. Critical traits of information
3. Information management
4. Fragile nature of information

These four areas will structure this discussion of information.

Defining Information

There are numerous terms that we use when dealing with information, including data, databases, and information itself. While many regard these three terms as synonymous, they are not; each has its own unique meaning. The starting point for information is data. Data, a company's building blocks, can include such items as representations, alphanumeric characters, pictures, drawings, or charts — in fact, anything that we can assign meaning to. Data can consist of numbers or qualitative input, and a quality problem can be described using both words and numbers. Both are important; neither can be ignored. With numbers, we would report the number of defective parts or the extent to which the performance of the process deviated from the standard. With words, we would enter any relevant observations, such as the fact that maintenance had recently been performed at the work center or that the parts being rejected were supplied by a relatively new vendor. Both the words and the numbers help to build a complete picture of the event — a picture that can be recalled at a later point in time.

Data are inputs that are ultimately stored in a structure known as a database, which makes the data available to everyone who needs the information. Data and the database do not necessarily have meaning on their own. Often they lack the structure or framework within which they should be presented. When we add structure or a framework to data, we have information. The structure provides context. The structure also focuses our attention by identifying what is important and disregarding the rest. It enables us to evaluate the data and draw inferences. For example, we have data in the form of a statistic of 50 units per hour. Notice that, without structure, this statistic has very little meaning. Now, when we add structure by stating that the standard rate of production is 75 units per hour, and there has been a recent change to design of the bill of materials/routing for this part, then we now have information. We know that the work area is producing at a rate below that expected (75 pieces per hour). We have also identified a potential root cause of this problem.

Managers use data as an input; they store data in a database, but they flag problems, identify issues, and resolve problems by using information.

Critical Traits of Information

When dealing with information, we must recognize that certain traits are associated with it. In most cases, we focus on three critical traits: accuracy, completeness, and currency. Anything pertaining to information, data, or databases should possess these three traits.

By accuracy, we mean that there should be a close correspondence between what we see with our eyes and what we record, enter, maintain, and use within an informational context. We have previously discussed accuracy within the context of inventory. That is, we have accuracy when the physical count equals the amount found in the computer system (given the allowance set by tolerance). Accuracy is critical because the various computer systems assume that the data found in the databases are accurate. Decisions are made and potential problems are flagged based on the quality of the data. If the data are wrong, then we can expect to see wrong decisions and actions being taken. At this point, it is important to remember that computer systems are no better than the quality of the data, databases, and information that drive them. This is why we have the famous dictum of GIGO (garbage in, garbage out). Accuracy occurs when changes that take place on the shop floor are reflected in the data in a timely fashion. If someone removes 10 units of a part, then that transaction must be captured and the data updated. Accuracy requires accountability and measurement. Somebody has to be held accountable for the accuracy of data. Without accountability, the only time that we have accuracy is after we physically check the entity or activity that the data represent.

The second trait, completeness, means that everything pertaining to the entity or the activity must be captured and recorded. For a bill of materials, completeness means that everything necessary for building that specific item is captured on the bill. For errors, completeness demands that all of the relative data are captured. This often means capturing not only the numerical evidence of the problem but also the qualitative/subjective observations of the users. Without completeness, we must recognize that we may be making decisions without having access to certain critical pieces of data. It could mean accepting an order when we do not have all of the material in stock.

Currency, the third attribute, means that the data, the database, and the information are up-to-date. This means that all of the transactions or changes to the data have been entered and the effects of these changes processed. Currency means that we have access to the inventory status as it exists now; it means having access to the latest version of the bill of materials.

Major Lessons Learned

- We are now working with manufacturing systems that are information dependent.
- When dealing with information, we must remember that data, databases, and information are related concepts but each has its own unique and specific meaning.
- Data must be both numeric and qualitative (i.e., data must include both numbers and words).
- Information is data to which structure has been added.
- All data, databases, and information must be accurate, complete, and current.

Managing Information

Overview

Having examined some of the fundamental concepts underlying information, we are now ready to discuss the remaining two points: (1) information management, and (2) the fragile nature of information.

Information Management

Information management deals with all of the activities necessary to collect, store, summarize, and retrieve or report information. Within information management, we integrate the three concepts of data, databases, and information.

The starting point for information management is the collection activity. It is here that we deal with such issues as:

- *What data should be collected?* This first issue is critical because it identifies not only what we will collect but, more importantly, what we will not collect. As we will show later, if we do not collect data about some activity or item, then that data can be viewed as lost for all intents and purposes. This is an important issue because, while we could collect data about everything, we do not want to do so. With too much data, we can encounter several problems: (1) the challenge of entering and storing all the data, (2) the possibility that the resulting databases may be too large for our systems to manage, and (3) "paralysis of analysis." That is, with so much data available, we may feel obligated to use it all (even if the value of this data is marginal at best). The challenge for management is to collect only the data needed — no more, if possible, and no less.

- *How frequently should the data be collected?* Here, we deal with the issue of frequency. The more frequently we collect data, the more transactions we have to process. Every transaction has to be entered, and every transaction creates some form of demand on capacity. In addition, the more frequently we collect data, the greater the possibility that we will collect data reflecting random events. This can generate a form of system nervousness — a state where we react to an aberration of data, rather than to a fundamental and persistent problem. However, we must weigh the costs of reducing the frequency of data collection against these factors. When data are collected too infrequently, we run the risk of missing or not capturing any fundamental and critical changes or problems. Every manager must pick that frequency that best balances the costs of collecting data too frequently and not collecting data frequently enough.

- *How do we collect the data?* This involves more than simply the challenge of whether we collect the data via a computer system or manually. Rather, this issue also deals with the type of data that will be collected. Specifically, what we are dealing with here is whether we will collect detailed, numerical data or whether we will simply collect "pass/fail" or "go/no go" results. The former generates more detail (and larger data files); the latter results in smaller files. Yet, with detailed data, we can do more subsequent analysis — something that we cannot do with simple attribute (pass/fail, go/no go) data. In addition, here we also deal with the issue of whether we will be limiting ourselves to numerical data or whether we will also include comments and observations.

■ *Who is responsible for the data?* For data to be accurate, complete, and current, somebody within the firm has to be held responsible for collecting, entering, and verifying the data. This responsibility must be assigned in the initial stage of the information management process.

The second stage in the information management process is that of storing the data. Here, we look at whether the data will be stored in manual form (e.g., Rolodex files) or in computer format. Further, if the data are stored in computer format, then we must deal with the specific mechanics of how it will be stored (e.g., as an independent file or as a file associated with a specific program). For most production and inventory control personnel, this aspect of information management, while important, is not as critical as the other stages.

In the third stage, we concern ourselves with how to summarize or keep the data. Data can be summarized either as individual observations or in terms of aggregate measures (e.g., means, standard deviations, total lead times). While storing individual observations can be very insightful, an important cost is paid, as there is now a far greater amount of data to manage. With individual observations, we can generate distributions of data, but with distributions we can see how changes in scheduling or order releases or setup time (to name a few factors) are influencing not only the means but also the shape of the distributions. As the shape changes, the predictability of events also changes. For example, a major element of the shape of a distribution is the range, or the difference between the lowest end of the distribution and the highest end. The larger the range, the less predictable is the outcome. To understand this better, consider the following situation.

We are faced with two work centers that can process the same part number. The first work center takes, on average, 5.5 hours to process the part number, while the second takes 4 hours to do the same task. Based on this information alone, we would pick the second work center as the preferred one; however, now we receive some more information. We find out, after collecting some data over time, that the range for the first work center runs from 5 to 6 hours. The same order on the second machine, though, can take anywhere from 2 to 16 hours. Now, based on this additional information, we change our mind. We decide to take the first work center as the preferred one. The reason for the decision is that the second work center, while apparently taking less time, is far less predictable. Ultimately, this lack of predictability translates into increased lead time, safety stock, and costs.

The fourth and final stage is that of retrieval and reporting. For most systems, this is the most challenging of the stages. While it is relatively easy to put information into a computer system, it is often far more difficult to easily and quickly retrieve that information. This means that we must develop a system that facilitates quick searching of the database and retrieval of data from it. As can be seen from this discussion, information management is something that should not be taken lightly within any manufacturing system. It is critical to the attainment of manufacturing excellence.

The Fragile Nature of Information

At first glance, it would not make sense to describe information as being a fragile resource. The term "fragile" would seem to be better applied to something like pottery, not information. Yet, information is fundamentally fragile because of its transitory nature. The data, if not recorded at the moment that they are generated, can and will become corrupted. Later on, if we want to capture the data, we have to rely on our memories. Details pertaining to the data are lost. If we cannot remember everything that has happened, then we either do not record the data or we create "data" that seem to make sense. Ultimately, failing to record data when the information is generated can compromise the integrity of the data, database, and information. And now you know why information is considered to be a fragile resource.

Major Lessons Learned

- Information management consists of four interrelated stages: data collection, data storage, data summarization, and data retrieval or information reporting.
- In data collection, we are concerned with such issues as what data we should collect, how frequently we should collect the data, how we should collect the data, and who is responsible for collecting the data.
- In data summarization, we focus on whether we should collect individual observations or pieces of data or aggregate measures such as means.
- Data retrieval is critical because we must recognize that one of the reasons that we collect data is so that we can subsequently use it. To use the data, we must be able to retrieve it quickly and efficiently.

The data are fragile, because if the information is not recorded as soon as it is created, then we run the very real risk of compromising its integrity. If we compromise its integrity, then the usefulness of that piece of data is lost.

When Is a Problem Not a Problem?

Overview

One of the most overlooked basics is that of management. For most people, management is decision-making. That is only part of basic management, though, as there is more to management than decision-making. Management is also problem-solving, but before we can understand problem-solving we must understand the critical differences that exist between problems, symptoms, and causes. To understand the importance of knowing these differences, consider the following situation.

Problem vs. Symptom

Recently, one of the authors had an occasion to work with a small manufacturing company. At one of the meetings, the author met with the firm's CEO, who gave the author a copy of his company's objectives and action plan. What the CEO wanted was an "honest" assessment of the plan (as it turned out, though, what the CEO really wanted was a blessing of the plan, not an assessment). In reviewing this document, the author's eyes came to rest on one item — "Reduce inventory by 25%." This is interesting, thought the author, and he circled it for future discussion. After reviewing the document, the author sat down with the CEO to discuss it. The author noted the interest in inventory. What did the CEO think were the underlying causes of the increase in inventory, asked the author. After all, we can never attack inventory directly, as it is a symptom of other problems. At this point, the CEO stopped and looked at the author with that expression reserved for small children and those people who simply do not seem to understand. He pointed out to the author that inventory was the problem, not the symptom. If the firm focused its energies, then he was sure that the inventory levels would drop. He also pointed out that he was surprised that the author did not seem to understand the difference.

This situation is not unusual. In many cases, we see people who are unable to differentiate between problems and symptoms. They often confuse the two. In many cases, they concentrate their attention on attacking symptoms, only to find that their efforts at improving the situation in one area have created a problem elsewhere in the firm. Problems and symptoms are related, but they are different. In this section, we will concentrate on these differences.

Symptoms

As pointed out in the above story, excess inventory is usually a symptom. Symptoms are often the first indicators that something is wrong; they are like a thermometer in that they can tell the manager that something is wrong but not *why* it is wrong. Typically, we can determine that we are dealing with a symptom by listening to the way that the concerns are described. When managers talk about symptoms, they use words such as "too much," "too little," or "not enough." For example, we will hear people talk about too much inventory or the scrap rate being too high or the output per employee being too low.

Fundamental to symptoms is the fact that they can never be attacked directly. Symptoms are residuals; that is, they are outcomes. To attack symptoms effectively, we must first understand the underlying causes — those factors that gave rise to these problems. If we attack a symptom, the result is often the "good news, bad news" syndrome, as described by Gene Woolsey of the Colorado School of Mines. For example, let us return to our previous inventory example. Attacking inventory levels without understanding the reason for their increase could, in fact, result in inventory reductions but at a cost. If the inventory exists because of processing problems (as indicated by high scrap levels), any reductions in inventory would result in shortened orders and missed due dates. The result would be that, without the protection offered by inventory, we would find that the number of parts ultimately produced would decrease. This would mean that we would be forced to ship incomplete orders or we would be forced to miss due dates. Note the "good news/bad news" aspect of this situation. The good news is that inventories have fallen. The bad news is that customer service levels have also fallen.

Ultimately, a manager's time is better spent dealing with problems and causes. This naturally leads to the matter of defining a problem. Problems are important and are the focal point of management attention. As Charles F. Kettering, one of the great innovators of twentieth-century business and a major player in the growth of General Motors, once noted, "A problem well stated is a problem half solved."

In general, a problem is a perceived gap. This gap may be between the present situation and some desired situation. For example, if the excess inventory in our example had been an indicator of problems in processing, then the problem statement might be stated as follows: "What can we do to improve the efficiency of current operations so that their processing performance is in line with other comparable operations in well-run systems?"

Urgency, Structure, and Orientation of Problems

When dealing with problems, we have to deal with three related issues. The first is the issue of *urgency*. Presenting a problem is an invitation to action. When we define a problem, we have defined a gap, and the implied response is that something should be done to close the gap. However, when we issue an invitation to action, we must recognize that action involves change, and many managers resist change. This is a fact of life. To define a problem successfully, we must also identify the impact statement, which asks the question: "What would happen if we did nothing about this problem?" It deals with the issue of living with the current situation. Urgency is created whenever we show that the costs of acting on the current problem are less than the costs of not acting. For example, there are many people who have stopped smoking because of the impact statement. Their doctors told them (often after a diagnosis of heart attack or other ailment) that if they continued to smoke they could expect to live about six more months, but if they stopped their life expectancy could be measured in years. It is important that we deal with urgency. There are many problems that are best left alone because the costs of change are far greater than any benefits that we could expect to gain.

The second issue is that of *structure*. Problems span a spectrum that ranges from being poorly structured at one extreme to well-structured at the other. Well-structured problems involve clear goals, well-understood means of achieving these goals, and complete and accurate information to identify and resolve the difficulty. When faced with well-structured problems, managers can often adapt and apply ready-made, routine solutions. For example, consider the situation of our car requiring an oil change. In this situation, we have a well-structured problem — to close the gap between the current state of the car and the desired state (a well-maintained and properly working car). The conditions triggering the need for an oil change are well defined (e.g., once every 3000 to 5000 miles). When this condition has been identified, the owner applies the routine solution of changing the oil to create the desired state — a well-maintained car. With poorly structured problems, managers lack good information. With an insufficient understanding of appropriate goals and the means to achieve them, they have trouble assessing the size of the gap between the current and desired state, or even whether any gap exists. In such situations, managers must assess current conditions, often gathering more information to do so. They must add structure to the poorly structured problem. This action must precede any further activity.

The third issue in regard to problems is that of *orientation*. Problems can be strategic, in that they deal with how the firm competes in the marketplace. They can also be operational, in that they deal with operational issues such as equipment breakdowns, absenteeism, and late deliveries. Operational problems have short time horizons, while strategic problems have very long time horizons. Why focus on the differences? The reason is that we can never really solve a strategic problem on an operational level or an operational problem at a strategic level. There are many problems encountered by managers that are really strategic in nature. Yet, they are seen as being operational.

Causes

Finally, we must deal with causes. The cause of a problem is anything that creates or contributes to a gap between the current and desired future situation. A cause is a source of observed symptoms. It is the condition that managers must identify and change to eliminate the symptoms. Causes and symptoms are closely linked. In many cases, several symptoms can be traced back to a few, common causes. On the other hand, different causes can create similar symptoms. Vague links between symptoms and causes complicate problem-solving, as managers must identify the most likely set of causes for observed symptoms, based on limited available information, and take appropriate action. We must recognize that under these conditions there is always a chance that managers may incorrectly identify causes and take inappropriate actions. This is a fact of problem-solving.

Major Lessons Learned

- Managers are basically problem-solvers. To be effective problem-solvers, we must understand the differences between symptoms, problems, and causes.
- Symptoms are indicators of problems, as they point toward difficulties. Symptoms are often suggested by the use of the word "too"— "too much inventory," "too little output," "too much scrap."
- Symptoms are residuals and cannot be attacked directly; rather, to eliminate a symptom, we must identify and eliminate the underlying causes.
- Problems refer to gaps between what we are experiencing and what we want to experience.
- When dealing with a problem, we must recognize the importance of urgency, how well-structured the problem is, and its orientation (strategic or operational).
- Problems should be addressed when the costs of change (or the benefits gained) are less than the costs of not changing. Identifying the costs of not changing requires the formulation and presentation of an impact statement.
- Causes refer to the sources of the observed symptoms which are what we act upon.

Understanding the Problem-Solving Process

Our previous discussion has explored some basic concepts of management and included an examination of such terms as problems, causes, effects, and symptoms. Now we want to examine the problem-solving process. In this section, we begin our discussion of this process. Before continuing, it is important to recognize that we are dealing with a process. Problem-solving is not an art; it is not necessarily something that can only be learned by going through the school of "hard knocks." It is not a skill that is given to some and not to others. Rather, it is a skill that can be taught. Central to the problem-solving process is recognizing that it consists of five major stages, and this discussion is meant to act as a checklist for that process.

Step 1: Size-Up/Description

For most readers, it would seem natural that the first stage in the problem-solving process is to formulate a problem statement; however, in many cases such an activity might be premature. Instead, we begin with structured information gathering in the size-up/description stage. There are several reasons why problem-solvers should begin the process by gathering information. First, the problem may not be clearly defined. Frequently, managers begin problem-solving knowing only that something is wrong. They have ample symptoms — late orders, rising inventories, falling quality, or an increase in customer complaints, but instead of reacting immediately, managers must develop a clear picture of the current situation and the desired situation.

Second, managers rely on the activities of this stage to confirm that they have identified the correct problem. One of the authors experienced the importance of this firsthand when he was called in to evaluate operations within a stockroom. The manager in charge of the stockroom had decided to implement a bar-coding system as a means for improving inventory accuracy. The bar-coding system, the manager argued, would improve inventory accuracy because it would simplify the recording of information. After studying the operation of the stockroom, the author noted that the need to reduce the amount of effort spent in recording information was an irrelevant argument. No one in the stockroom bothered to write anything down. They just took what they needed and walked out (after noting that they would write down the information later on, which meant never). The first step was not to install a bar-coding system; rather, it was to instill some sense of system discipline. At this point, the manager dismissed the author, noting sarcastically that it would be nice if the author would ever learn to assess operations correctly. It was at this moment that the author uncovered the real problem — the manager wanted a bar-coding system because he saw one operate during a tour of another plant.

Third, we need a size-up/description stage to ensure that we are working on problems that we should solve, rather than working on problems that we think we can solve. This situation often occurs when managers become proficient users of certain types of tools (e.g., spreadsheets, computer simulations). The authors saw one manager who used ABC analysis for every problem, the reason being that it was the only tool that he really understood.

Finally, a size-up can reveal new problems that others in the firm have overlooked. This occurs because people tend to experience the same situations every day. As a result, they accept a particular situation as being normal, rather than seeing it as a problem.

Information gathered during the size-up/description stage includes:

- *Company status:* An examination of the firm, its products, its financial health, and its methods for selling itself and its products in the market-place.
- *Customers:* Information about who buys the products, how they place orders within the system, and what they expect (i.e., what the customer considers to be value). This review may uncover that the firm is not serving just one market but many different markets, each with its own desires and requirements.
- *Order winners, order qualifiers, and order losers:* A determination of what we must do better than anyone else (the order winners), what we

must do good enough (the order qualifiers), and what we must avoid doing if we do not want to lose orders (the order losers).

■ *Critical system tasks:* Identification of those tasks at which the firm must excel if it is to succeed. These tasks should be consistent with the order winners, qualifiers, and losers.

Typically, during the size-up/description stage, we might ask the following questions:

■ What is happening?
■ What are the major indications of a problem?
■ Who is involved in the situation?
■ What does the firm really provide to its customers (what does it sell to its customers)?
■ Who are the customers?
■ What do the customers really want from the firm and its production and inventory control system?
■ What does the firm have to do better than its competitors to win orders from customers?
■ What does it have to do as well as its competitors to be considered as a supplier?
■ What are the critical tasks that the production and inventory control system must achieve?
■ What unique or special features are evident in customers, processes, or products?
■ So what?

Managers should collect this information to uncover issues and facts that affect their formulation of a problem. Asking the "so what?" question can help to focus the investigation. If the answer does not reveal anything useful, then the subject of the question is not important and it should be dropped. The goal of this stage is to uncover the critical issues and to make sure that these issues are highly visible within the problem-statement stage.

Step 2: Problem-Statement/Diagnostics

The next stage is to state the problems that require attention. In this stage, the manager states the current situation and lists the major symptoms, causes, and triggering events. Also included in this stage should be identification of the desired outcomes, as well as development of the impact statement. The impact statement is a critical part of this stage, as it explicitly describes what can happen if this problem is not addressed. This statement is needed to overcome the management inertia often encountered in any firm. It is intended to show people that the costs of changing or addressing the problem are less than the costs of ignoring the problem.

Step 3: Analysis

The third problem-solving step, analysis, often overlaps with the problem-statement/diagnostics stage. As its major task, analysis explains the development of the current situation. It details the relationships between features of that situation and outlines its problems, causes, and symptoms. By defining causal relationships, the analysis stage directs attention to the changes needed to solve the problems described in the problem-statement/diagnostics step. This forms a foundation on which to construct strategy alternatives in the following prescription/alternatives step.

Analysis often asks questions such as:

- What has happened?
- Why did it happen?
- What sequence of events caused the observed problems and symptoms?
- What key factors have influenced these events?
- How can managers change these factors?

Major Lessons Learned

We will stop at this point to review what we have learned thus far and will resume our discussion of the problem-solving process below. Up to this point, we have introduced the concept that problem-solving is a process, and we have raised the following important points:

- The process of problem-solving consists of five well-defined steps (three of which we have covered). In laying out these steps, we are trying to present a checklist of sorts. It is often a useful idea to go through these five steps to ensure that we have not omitted any important areas or issues.
- The first step in the process is the size-up. This step accomplishes something very important — it is a check to make sure that we have enough information and that we are dealing with a real problem (rather than a symptom or an artifact).
- The next step is the problem-statement/diagnostics. It is here that we identify the symptoms, the underlying problems, and the impact of the problems. It is also here that we explicitly address the issue of why we should address a particular problem through the impact statement. The impact statement describes the results that can be expected from ignoring the problem.
- The third step is analysis, which serves the useful purpose of describing how we got into the current predicament. This stage helps us to explain what happened. This understanding helps direct our attention toward identifying possible areas to explore and issues to consider.

In the preceding discussion, we focused on understanding the problem-solving process. As we pointed out, problem-solving is not an art but rather a well-structured process that consists of five steps, each of which involves a number of concerns and issues that should be addressed either explicitly or implicitly. We will now finish our discussion of the process by looking at the remaining two steps.

Step 4: Prescription/Alternatives

In the fourth step, prescription/alternatives, the problem-solving process focuses attention on developing a solution to the problem that was evaluated during the size-up/description stage, formally stated in the problem-statement/ diagnostics stage, and explained in the analysis stage. Based on the results of analysis, managers identify problem features that they must attack or change to close the gap between the current and desired future situations. They also establish criteria for an acceptable solution, specifying time limits (e.g., any solution must take no longer than 18 months to implement), cost limits (e.g., any solution must provide a rate of return greater than 15%), and strategic emphasis (e.g., any solutions must reinforce the firm's position as the quality leader in its industry).

This step should not try to present a single solution, but rather a number of different alternatives. Managers should present each alternative in a brief statement containing the changes being proposed, why and how those changes should improve operations, and the strengths and weaknesses of that particular alternative. This step frequently asks such questions as:

- What conditions must a solution alternative satisfy?
- What are the major features of each proposed alternative?
- How would each alternative improve operations and address the problem?
- What are the strengths and weaknesses of each alternative?
- Does the alternative propose a realistic solution to the problem?
- Does the alternative address all of the problem features and causes identified in the analysis step? If not, why not?

It is important to never underestimate the importance of this stage. Often, one of the most important outcomes that a problem-solver or manager can provide is that of proposing alternatives. These identify a range of actions to be taken to address a problem or issue. In many cases, the ultimate alternative selected may not be the one that we may have originally recommended but instead might be one of the alternatives identified in this section. In the alternatives section, the value that you as a problem-solver provide is that of generating different options — options that you see in your role of problem-solver.

Step 5: Implementation

In the final step, implementation, the problem-solving process formulates a single solution and puts it into practice. The previous four steps have carefully explored the current and desired future situations and developed a solution that will effectively close the gap between them. Implementation involves choosing one of the alternatives identified in the previous step or some combination of those alternatives when one cannot solve the problem by itself.

After selecting a solution alternative, managers must develop an implementation plan that describes how they intend to put it into practice. This plan specifies the changes that managers intend to make and the order in which they intend to introduce those changes for both short- and long-term actions. The plan should describe short-term actions to improve the effects of the problem and long-term actions to eliminate or control its causes.

This step does not end when managers present an implementation plan, though. An effective response requires actual changes in company operations to implement the chosen alternative successfully. Much of this success depends on how completely people affected by the solution accept its changes. A solution that makes sense to managers may not necessarily gain acceptance by the people on the shop floor, especially if that solution will eliminate some of their jobs. We have identified four factors that are instrumental in securing acceptance of solutions (especially ones that are potentially painful to accept and implement):

1. *Survival:* We can really get people's attention when they know that the only alternative to our proposed plan of action would result in the loss of the firm, group, or department (along with the resulting loss of everyone's jobs). For example, one company's managers proposed to automate operations, which would reduce its employment from 500 to 350. To sell the idea to workers, the managers showed that the current system could offer 500 jobs, but only temporarily as it could not keep the company competitive and it would ultimately fail. Automation would offer fewer jobs, but more permanent ones, as it would allow the company to be far more competitive. The employees agreed to accept the change to automation. Remember the old saying, "Nothing so focuses the mind as a hanging."

2. *Lack of surprises:* Employees tend to accept solutions that eliminate nasty surprises that the current system causes. Remember the simple fact that in business few, if any, surprises are pleasant. For example, one of our pieces of equipment has broken down — surprise!

3. *Simplicity:* People tend to accept anything or any action that makes their lives and their jobs easier to do.

4. *Pride:* Most employees are fundamentally driven by the desire to do "good work." What prevents them from achieving this objective is often the system. If we can provide recommendations that help them do their jobs better, they will often accept these actions (provided that we recognize their roles and properly reward their involvement).

When dealing with implementation, there are several other issues that we should always remember:

- The implementation plan should be feasible. We should be able to implement the recommended actions given current or anticipated resource levels.
- The implementation plan should be consistent. Set decision-making criteria (reduce lead time, cut costs, improve quality) and develop a plan consistent with the criteria.
- The plan should cover both the short term and the long term. Most tend to focus on one of these two, but not both. The short term is highly operational and has a strong sense of immediacy. In it, we should be able to answer the following question: "What is the first thing that you would do upon coming in on Monday morning at 8:00 a.m.?" That is, we should have identified some very specific actions to be carried out. It is not enough to say that we would hire a consultant or call a department or team meeting. We must be prepared to do something; however, we should also never forget about the long term. At times, some of the recommendations that we make in the short term are designed to stop the "bleeding." That is, they are intended to address some critical issues. However, they do not eliminate the problems from reoccurring because they have not addressed the underlying root causes. Addressing these root causes may require a long-term perspective. For example, in the short term, we may use a new scheduling system or allocation system to deal with a situation where demand greatly exceeds capacity, while in the long term we must address the lack of capacity. As Keynes, the great economist, once noted, "I think about the long term constantly because that is where I am going to spend the rest of my life."
- The plan should be prioritized. When stating our plan of implementation, we should think in terms of priorities. That is, we should state the important issues and recommendations first. These should always be the starting point.
- We should think in terms of ABC analysis (which we discuss as it applies to inventory in Chapter 8). In this context, the A items are the most valuable, and the C items are the least. The point of the discussion

is that we should focus our attention on the A items, which are few in number but account for a large percentage of the inventory value. Similarly, when it comes to the implementation, we must identify our A items and focus on them. This is where we should spend the bulk of our time and effort. These are the critical items. They are few in number but essential. After we dispose of these items, then we would go on to the B items. Finally, if we have the time and space, we can touch (and only touch) on the C items. Beware of the illness that affects many managers. They have spent a great deal of time and effort collecting data, analyzing information, and making sense of what is going on. Early on, they have been able to identify the A items, but after expending so much time and effort, they look at these items and react with horror. These items seem so obvious that surely they could not be the critical issues. As a result, they are largely ignored while time and space are devoted to the less obvious B and C items. As a result, the report has lost its effectiveness. Remember, what may be important but obvious to us may not be so to someone else. We must focus on the vital few and ignore the trivial many.

■ We should present our plan using the recommended format. A final small point. If the boss wants us to use a specific format, we will use it (even if we think it makes as much sense as Etruscan, a dead language that no one has yet been able to translate). Formats have been developed to simplify the reading and comprehension of re-ports. If we do not use them, we will only succeed in upsetting our reader.

General Observations About the Problem-Solving Process

Effective problem-solving requires careful attention to each of the five steps in this process in order. The sequence of the steps is important because it trans-forms the process into a sort of checklist that helps managers to avoid ignoring or overlooking valuable information. Inexperienced problem-solvers often try to move straight to the problem statement in an attempt to quickly frame a decisive solution. This haste leads to spending too little time in the critical size-up/description stage or even skipping it entirely. Time invested in sizing up a problem often pays returns in improved problem statements and more thor-ough understanding of a problem's background, which can help to smooth the implementation step. The five-step problem-solving process imposes some order on a potentially chaotic situation by providing a structure for under-standing and responding to that situation. We have provided this discussion because it has been our experience that few managers really have a good mastery over the problem-solving process.

Responding to a Problem:
Solve, Resolve, or Dissolve

Overview

When dealing with problem-solving, it is naturally assumed that the goal is to solve the problem. That is, we want to identify and implement the best (optimal) solution to the problem. There is something magical about the notion of a best or optimal solution. This denotes something that we cannot improve upon. However, if we were to focus on optimal solutions, we would find ourselves greatly limiting the range of problem-solving stances available. We may also find that there are other approaches that are far more effective and efficient.

Identifying the Three Forms of a Solution

Ultimately, there are three basic types of solutions that we can solve and implement in response to a problem — we can solve the problem, we can resolve the problem, or we can dissolve the problem. Each takes a very different tack; each brings with it its own unique approach and cost/benefit trade-offs. To understand the differences between these three approaches, we will compare them by applying them to the problem of inventory management. Suppose we are faced with a situation in which we need to identify the most appropriate order quantity for a particular product. The product is made to stock (so we must have it on hand). Furthermore, the demand for the product is fairly stable and regular at 1200 units per month, with very little seasonal variation. It costs the firm $1 to keep one unit of the product in stock for one month. The cost of equipment setup to make a batch of the product is fairly high and has been assessed by accounting at $150 per run (this is a true setup cost in that it is actually incurred every time we run the product through the equipment). Furthermore, capacity is not an issue for this product. Finally, the cost per unit is not influenced by the size of the order quantity. With this information, we can now look at the three approaches and the types of solutions that they would create.

With the *solve* approach, we are interested in determining the best or optimal solution for this set of conditions. A manager using this type of approach might turn to the economic order quantity (EOQ) to determine the most appropriate order quantity. The EOQ formula tries to determine the order quantity that generates the lowest total cost (where total cost is limited to holding costs and setup costs). The logic used by the EOQ dictates that this point is reached when the total holding cost equals the total setup cost. With this approach, the manager might take the information found in our problem, plug it into the EOQ, and find out that the lowest cost point occurs at 600 units (for those of you who know the EOQ, you can verify this calculation). With this order quantity, the manager who has solved the problem can argue that there is no other order quantity that would generate as low a total cost.

The second approach is that of *resolve*. With this approach to problem-solving, the goal is not to get the best solution but to get a "good enough" solution. This solution often reflects considerations frequently ignored in the solution approach. For example, another manager reviewing the preceding solution might note that this order quantity would not make sense, as the standard container size used within the plant is that of 575 units. Implementing the EOQ/solve solution would mean that each batch would require two standard containers. The first would be completely filled with 575 parts, while the second would be only partially filled. It would make greater sense to set the order quantity at 575 units. After all, this manager might argue, the cost penalty paid for such a small deviation is relatively small (as we get close to the minimum cost point identified by the EOQ, the cost curve tends to bottom out, with the result being that the differences between points become very small). In addition, the benefits gained from the continued use of the standard container size would more than outweigh any such cost penalties. The order quantity of 575 units, while not optimal, is nevertheless "good enough."

The final approach, the *dissolve* approach, is in many ways the most radical and, ultimately, most effective. A manager using this approach might begin by asking why the setup costs are so high. After all, if the setup costs were relatively low, then it might be possible to go for an approach of ordering no more than is needed. If we need a batch of 5, then it would make sense to make only those 5 units. If we respond with a batch of 600 or 575 units, then we would use the 5 and put the remaining units into inventory. If we do not use the items immediately, then we have to deal with the problems of storage, uncertain demand, and obsolescence, perishability, and pilferage. However, we know that we are prevented from making only what we need by the high setup costs. To dissolve the problem, we must reduce the setup costs. If we can reduce the setup costs, then we would no longer have a need for inventory. If we want to reduce setup costs, then we must study these costs and determine why they exist and what can be done to eliminate or greatly reduce them.

As can be seen from this discussion, the dissolve approach is radical because it attempts to eliminate the problem from occurring in the first place. We see numerous examples of the dissolve approach being applied. A firm finds that it does not have expertise in making and storing one specific type of component. As a result, the company turns over the manufacture and storage of that type of item to a vendor who is an expert in the product. In another example, we see that we have a great deal of inventory that is tied up in the logistics pipeline between our vendor (who is located some 2500 miles away from us) and ourselves. Our dissolve solution to this problem is to encourage the vendor to locate a facility very close to ours. Ideally, we would like to see co-location (where the vendor is physically located next to us).

Before leaving this discussion, it is important to note that, as a result of developments such as Just-in-Time manufacturing and Total Quality Management, managers are now recognizing that the most useful approach to a long-term solution is that of dissolving. The other two approaches, we now see, are simply ways of coping with the problem in the short term. They make the problem less onerous, but they ultimately do nothing about removing the problem and its underlying causes.

To Go Forward, Sometimes We Have To Go Backwards

Overview

Imagine that you receive a telephone call from a friend who is on his way to visit you. That person is calling to tell you that he is lost. You ask him for any landmarks, but he can't give you any. You then ask him to tell you the route that he took to get to his current position, but he can't seem to remember. At that point, you inform your friend that: (1) he is really lost, and (2) without any information about how he got to his current location, you can't help him. What we have in this situation is an example of the failure of someone to understand exactly how they got into their current state, a situation often encountered in manufacturing.

There is one situation that we frequently encounter when we are out in the field. We go into a company where there is a crisis (everyone is aware of it), and we are approached by a manager who describes the situation, the symptoms, and what he thinks the underlying problems are. Typically the next thing that he asks for is a solution for this crisis. In that manager's mind, this is the correct thing to do. He knows that he is faced by a crisis, he knows that it is creating problems for him, and he knows that the crisis has to get resolved immediately. Yet, often this attitude of "let's concentrate on getting the problem solved right now and we'll think about what happened later on" is potentially dangerous. The reason is that this manager is overlooking some very important items of information. The manager has forgotten that before we can solve a problem, we must first understand how the firm got itself into its current crisis. By ignoring the past we are overlooking several key items of information — all necessary to prevent ourselves from repeating the same errors or to uncover potentially fatal flaws in the firm's operations. Furthermore, if we do not go back, then eventually what we end up with is a stack of Band-Aids — one on top of the other.

The Importance of Going Back

Why reexamine the past? The best way of illustrating the importance of going back is to look at a situation encountered by one of the authors. There is a company in western Michigan that is now recognized as one of the leaders in the introduction and implementation of Advanced Manufacturing Technology (AMT). This firm has developed a reputation for being able to generate predictable and consistent lead times; however, it was not always this way. The management had become aware of the fact that their processes seemed to generate operation lead times that were highly variable.

The management of the firm assigned some of its engineers to study the problem. These engineers studied the operation in each of the three shifts (this firm ran three shifts a day, seven days a week). What they noticed was that on every shift the processing times for this specific operation were remarkably consistent. However, what was interesting was that when they compared the processing times across the three shifts, they found that they were significantly different. When these processing times were collected and displayed in the form of a histogram, the result was variance. This led to the natural question of "Why?" What the engineers found when they pursued this problem was that different people on each of the three shifts had identified "process improvements" that were implemented during that particular shift by the people who had developed them (without any prior approval of management). These improvements, as you can expect, differed from shift to shift. Whenever a person was promoted, it was common practice for that person to teach his successor the exact same way to perform the improved operation. Eventually, the replacements would identify their own improvements and the cycle would be repeated. As a result, over time the processing times (and the processes that generated them) began to differ.

What the engineers and the managers had uncovered was more than simply a problem in processing-time variances; rather, it was a more basic problem of process control and managing process improvements. Had they simply focused on the differences in processing times (and not gone back to understand what had caused them), the result would have been a short-term improvement followed by a return to the same problem over the long term. Because they had bothered to study the past, the managers had developed a different solution — a solution that focused on standardization and controlling the manner in which changes are introduced.

Today at this firm, there is only one way of doing that particular process. Everyone, regardless of the shift worked, must carry out that operation in the same way. If someone on a shift uncovers a better way of doing things, then he or she must first show one of the managers this "new and improved" method. If it is truly an improvement, then this new procedure is made the standard for every shift. By looking at the past (by going backwards), management had learned that they were actually dealing with a symptom and not the problem. By understanding what was really happening, they were able to effectively address the underlying cause.

How To Go Backwards

When a crisis occurs, we must realize that there are two things we must do. The first thing that we must do is to stop the "bleeding." That is, we must do something now to stabilize the situation. This can be as simple as subcontracting or adding overtime. The second thing that we must do is to conduct a postmortem. That is, we must study the chain of events that led up to the current crisis, and we must understand what caused the various links to be put in that chain. The analogy of a chain is highly appropriate. Few crises ever occur because of a single action; instead, they are the result of a series of actions. Often, these actions build on each other. In many cases, if this chain would have been broken at any point, the crisis that we are currently observing would not have taken place.

Understanding these actions is important for several reasons. First, past actions may have resulted in constraints that can affect what we can or cannot do now. Second, by understanding what happened, we can often attack the underlying causes. Third, we can identify the critical players — those people whose involvement in the process of correcting the underlying problems is critical for the successful implementation of any proposed solution. These are the people who have to develop "ownership" in the solutions before these solutions can work.

Going backwards is a discipline. It begins by understanding what has happened. The question that we are trying to answer is that of how we got into the current situation. Answering this question means talking with the people involved, and it also means trying to convince these people that we are not trying to assign the blame to someone (it is important to remember that often the innocent are the ones who get nailed). Our goal is to uncover where the flaws are in the current system. In this respect, we are applying Juran's Rule: "When an error takes place, 85% of the time it is the fault of the system; 15% of the time it is the fault of someone involved in the process."

At this stage, we can draw on a number of tools to help us reconstruct the problem. These include procedures such as the "five whys" (i.e., ask "why?" five times; by the time we get to the fifth "why?" we will either find out that the person does not know what is going on or we will uncover the real reasons for the problem or behavior). The tools also include quality procedures such as cause-and-effect analysis. With the information that we have gained during this postmortem, we can now develop a solution that is aimed at preventing the problem from occurring again. This is the ultimate objective of any problem-solving exercise.

Major Lessons Learned

From our discussion, we have learned the importance of understanding what happened in the past and how these past decisions have influenced our current position. We have also shown that by reconstructing the events of the past we gain a great deal of insight — insight into potential underlying problems, insight into potential constraints attributable to decisions made in the past, and insight into potentially dangerous decision-making patterns. We have shown that if we want to avoid applying a series of meaningless Band-Aids, then it is important that we spend our time understanding the past. We have shown that to go forward, there are indeed times when we must go backwards.

3 What Type of Company Are We?

Why Should We Care What Type of Company We Are?

Most firms know that they build products and deliver services. They recognize that they embody some form of manufacturing or transformation process. However, when questioned, most managers would have difficulty in describing what type of firm or process that they have. In fact, if we were to ask the manager of a "typical" firm to describe their process(es) and to identify the extent to which they are similar to other systems, we would find that they would see themselves as being unique. While it is true that every company is, to some extent, unique, it is also and more importantly true that every company's manufacturing/transformation processes can be categorized as belonging to one of a finite set of possible systems.

This commonality is critical because it allows managers to learn from the experiences of others. It is also critical because it helps us to develop expectations of how a specific system should perform, what its bottlenecks are, and the type of flows, type of equipment, and type of planning and control systems that we should expect to see. We can then use these expectations to evaluate systems and to look for gaps between what we expect to see and what we actually observe. These gaps often flag opportunities for improvements. Further, this commonality is important because it helps to guide our selection and evaluation of potential software packages.

Given the importance of this commonality, it is important that managers be able to assess their systems and determine which of the various generic systems really best describes or captures their current system. Before they can carry out this assessment, managers must first know the types of basic generic systems that are out there and those traits that differentiate one type of system from another. Developing this type of understanding is the primary objective of this major chapter of the book.

What Type of Company Are We? The Major Dimensions

Overview

We now address a question that is frequently asked of us: "What type of company are we?" To understand this question, we must first understand the origins of the question. As most readers of the APICS journal are aware, the journal provides software directories on a regular basis. These directories contain listings of various software packages according to such categories as MRP/ MRPII, Finite Capacity Scheduling and Planning, Warehousing, Forecasting, and Simulation. These directories not only identify the various software packages and their vendors, but they also detail various features of the software. One of the categories pertains to the specific manufacturing settings for which the various packages are potentially suitable, such as job shop, batch, repetitive, process/continuous, mixed mode, medical/drugs, food, and automotive. For many readers, these categories are a source of major confusion. They are not clear as to the differences among the various categories. In this section, we will explore these various categories of manufacturing settings and their implications. First we will focus on the categories of job shop, batch, repetitive, and process/continuous.

Manufacturing Settings: Major Dimensions

Before we begin to explore the various manufacturing settings, it is important that we identify several of the major dimensions underlying these various settings, such as:

- *Volume* refers to the level of output that the setting is capable of economically supporting and consists of two levels. The first is that of frequency, or the number of times during a period of time that we can expect to produce a lot of a specific part (i.e., a specific part number or stockkeeping unit). The second is that of repetition, which refers to the size of the typical production order. That is, whenever we release an order for a specific part number, repetition refers to the size of the order. Using these levels, we can see that low volume can be generated by a combination of low repetition and low frequency, or very low repetition and moderate frequency, or moderate repetition and very low frequency. Volume determines the size of runs that the manufacturing setting can be expected to deal with.

- *Variety* refers to the extent to which the products built are standard or subject to modification. These variations show up in differences in routings that the manufacturing setting is expected to accommodate.
- *Nature of flows* refers to the extent to which there is a dominant flow through the manufacturing system. In some conditions, we must accommodate a setting in which every order released to the floor has its own unique routing. In other conditions, nearly every order follows the same routing.
- *Equipment type* refers to the type of equipment typically found in the specific manufacturing setting. Generally, this dimension runs from general purpose (GP) equipment (e.g., a drill, grinder, deburrer) to special purpose equipment.

In addition to these dimensions, it is also important to note that not every manufacturing setting is "pure." That is, we seldom see a pure job shop or a pure assembly line. Rather, we see a job shop that has elements of an assembly line in it. Under these conditions, we tend to describe ourselves in terms of that manufacturing portion that is either most critical or the largest.

Manufacturing Settings Examined

Now we will examine four types of manufacturing settings: job shop, batch, assembly line, and process/continuous.

Job Shop

The job shop anchors one end of the manufacturing spectrum, with process/ continuous being the other end of this spectrum. This setting is characterized by very low volumes, with the production runs consisting typically of between 2 and 50 units. In addition, variety is very high in this setting (a result of the small runs), with each order being very different from the others in this system. As a result, software packages targeted toward this setting are designed to cope with this high level of product variety. The combination of low volume and great product variety results in a great deal of diversity in product routings. This diversity places tremendous demand on the ability of management (and the supporting software package) to manage the flow of work through the shop. This means that dispatching at the work center level is a critical task.

In general, the equipment in a job shop tends to be general purpose in nature. This means that, while setups are relatively short, the processing times per unit tend to be longer than what one would find in an assembly line. This short setup time is consistent with the demands for flexibility made by the orders in this setting. Although the equipment is general, the skill level of the employees tends to be very high. In most job shops, the key bottleneck tends to be labor, not equipment.

Layout in a job shop falls into one of two formats. The first format is *functional.* In this format, we design the flow so that all of the same or similar equipment is placed together in the same area. In contrast, the second format of *process* organizes and locates together all of the equipment needed to build orders belonging to the same part family (more about this trait later on in our discussion of batch). In general, most of the lead time in a job shop is consumed by queue time (i.e., orders sitting at various work centers waiting to be processed).

Batch

With batch, the volume is increased, and the typical run quantity ranges from 50 to 2000 units. Variety, while still present, is less extensive than that found in the job shop. With the larger runs and less variety, we now find less general purpose equipment and more specialized equipment and tooling. An important feature of a batch environment is that of the part family and machine cells. A part family consists of a set of parts which share commonalties or similarities in terms of processing (the most important feature from our perspective) or design. With part families, we must now recognize the presence of sequence-dependent scheduling. That is, the setup time depends on the order in which we group the runs. The ability of a software package to recognize and incorporate this feature into the resulting schedule is an important consideration. Typically, Material Requirements Planning (MRP) systems are most appropriate for this setting.

Repetitive

With the repetitive setting, we move into a high-volume manufacturing environment where the units of output are discrete (e.g., tires as compared to gallons of paint). Here the items are either standard or they are assembled from standard components and options. The presence of this standardization changes the requirements placed on the planning and control system. With such large volumes, we now find ourselves with dominant routings. That is, most of the orders follow a set sequence of operations (the machines and work centers are physically laid out to correspond to this flow). Because of the volumes, we have to focus our scheduling activities at two points — the master schedule and the first operation (frequently called the gateway).

Because of the large volumes combined with the very short processing lead times, we tend to manage inventories in a very different fashion. In the previous systems, we could track the inventory by focusing at the points in time when inventory was issued to the orders. In many repetitive systems, we use a backflushing logic. That is, we determine the number of end items we have produced during a period of time. We then break this level of output into its components. Ideally, we should arrive at the ending inventory by taking the beginning inventory, adding to it the receipts received during the time period, and then subtracting from it the number of components needed to cover production (as based on the number of end items produced) and scrap allowances.

Equipment and tooling capacity within the repetitive environment tends to be specialized (again, due to the high levels of volume of standard items), with labor being either unskilled or semiskilled. Within this environment, it is equipment and tooling capacity that tends to be the bottleneck, not labor. Capacity planning is very critical within the repetitive environment; therefore, attention must be paid to the strength of the capacity planning modules.

Process

Process is very similar to repetitive except that the units of output are not discrete but continuous. Oil and paint are examples of items produced within a process environment. Here, volume is high, and the products are fairly standardized. In addition, there exists a dominant flow through which every product moves. Because setups (or the changes required to go from one product to another) are fairly time consuming and expensive, care must be taken to ensure that setup considerations are incorporated into the software package and the planning and control system. Typically, we tend to see cyclical production scheduling. Because of the high volumes, process settings are characterized by high levels of capacity utilization. In this type of environment, planning is the critical activity. Execution tends to be a reflection of planning.

Major Lessons Learned

- There should be a strong fit between the manufacturing setting and how the system is managed and controlled.
- The settings are differentiated by factors such as volume, variety, nature of flows, and equipment type.
- The environments fall along a spectrum anchored by job shop at one end and process at the other.
- Each environment has its own requirements that must be recognized within the software decision.

A mixed environment exhibits traits of different settings; yet, there is a critical or dominant setting which can be used to identify the targeted setting if there is no package supporting a particular mixed-mode production.

What Type of Company Are We? Understanding the Industrial and Manufacturing Contexts

Overview

Previously, we looked at the question of "What type of company are we?" This question resulted from a frequently asked question that stems from the various surveys and software directories compiled by *APICS: The Performance Advantage*. As we have discussed, these directories provide readers with the following choices for situations that the software packages address:

- Job shop
- Repetitive
- Aerospace/defense
- Food processing/food industries
- Medical applications/hospitals
- Mixed-mode (combination of job shop, cellular, process, repetitive)
- Drug industry
- Automotive
- Process

In our previous discussion, we focused on the job shop, batch, repetitive, and process categories. These can be best described as pertaining to the *manufacturing context*. That is, our interest is in the manufacturing process and its underlying structure and inherent traits. As we explained, these four categories are assessed along the dimensions of volume, variety, nature of flows, and equipment type; however, we did not really address the remaining categories, which are fundamentally different in that they refer to the *industrial context*.

Understanding the Industrial Context

The industrial context involves requirements and constraints placed on firms operating in these industries. Typically, these constraints and requirements reflect practices such as:

- *Order release and scheduling:* In some industries, firms are required to handle certain types of order releases. For example, in the automotive industry, the Big Three have developed a specific type of order release system that everyone is expected to accommodate.

■ *Lot and batch tracking:* In other industries, such as drug and medical, lot and batch tracking is very critical. Firms are expected to be able to identify a particular item in terms of specific production batches and order lots.

■ *Information storage:* Again, in some industries (specifically, those that are either regulated by or work with governmental agencies and groups), a requirement for detailed and specific data storage, management, and retrieval is present. These requirements must be satisfied by the software package selected for this environment.

■ *Electric and/or communication interfaces:* Some industries require that firms must be able to handle or accommodate certain types of communication and/or electronic interfaces. For example, the ability to utilize electronic data interchange (EDI) or bar-coding is an example of the technology that must be accommodated. It is well known in the automotive industry that suppliers wishing to deal with the major manufacturers must be EDI compliant.

■ *Inventory tracking and control:* In some industries, such as food, we must pay a great deal of attention to the tracking and management of inventory with the goal of reducing or eliminating the problem of spoilage or pilferage. Some items within these industries have a great deal of value on the street. This point was brought home to us when one of the authors was doing a presentation for the local NAPM chapter located in Kalamazoo, MI. Before the presentation, the author spoke with a purchaser who worked at a large pharmaceutical company and mentioned the various hurdles that he had to overcome because of a catalyst that he was using to test the quality of a certain drug. He said that the process required between 5 and 10 pounds of this substance. Before he could get the product, he had to sign out for it, and he had to have the substance checked before he could dispose of it. All of this for 10 pounds of "speed" (the street name for the catalyst).

These are some but not all of the dimensions of the industrial context.

Understanding the Impact of the Industrial Context

This discussion brings us to an interesting question — "So what?" Why is it important that we recognize the industrial context as a separate dimension? The reason for breaking out the industrial context is that this dimension can and does operate independently of the manufacturing context. As a result, it is possible to have a firm operating in the automotive industry in need of software that can meet the needs of the industrial context, but it is also possible for that same firm to have multiple, different manufacturing contexts (e.g., job shop, repetitive, or process). The result is a matrix with industrial context on one side and manufacturing context on the other.

How Does This Information Help with Software Selection?

With this information, we can now see that the task of selecting the "right" software is slightly more complex than we had previously thought. To pick the "right" software, we must pick the software package that has features consistent with both our manufacturing context and our industrial context. The two dimensions must be jointly considered; however, in many cases, we can downplay the importance of the industrial context. There are instances in which the industrial context is not as important. The industrial context issue is important when dealing with situations pertaining to the industries previously mentioned, such as aerospace/defense, food processing/food, medical applications/hospitals, drug, and automotive. Outside of these industries, however, the manufacturing context becomes the dominant one.

Major Lessons Learned

- When looking at software packages, we must consider the issues of manufacturing context and industrial context.
- Industrial context deals with requirements and constraints imposed by specific industrial settings. Typically, these settings pertain to issues of order release/seclusion, lot and batch tracking, information storage, electronic/communications interface, and inventory tracking and control.
- There are many environments in which the industrial context can be downplayed.
- When picking a software package, we must select that package that meets the needs of both the industrial and manufacturing contexts.

4 Understanding the Importance of Metrics

Why Metrics?

If you had tried some five years ago to develop a list of topics that form the basics of operations management, you probably would never have included the topic of metrics. Until recently, metrics, the process for capturing, measuring, reporting, and assessing the performance of activities, were largely overlooked. As a result, this process was often considered after the fact ("now that we have the system in place, how do we measure and report its performance?") and often was carried out by other groups (typically accounting). It was also an activity that was assumed to be more punitive than corrective. After all, measurement being done after the fact often meant that the factors that created the problem in the first place had passed. Because they were history, there was little opportunity for correction.

In the last five years, however, there has been a significant increase in interest in metrics. This interest is a result of several factors. First is a new awareness of the costs created by poor metrics. With poorly thought out and poorly integrated metrics, we find that there is often a great deal of confusion on the part of the users. In addition, managers have encountered numerous situations in which the metrics developed and used within one area ran counter to the metrics being used by another group or area. The result of this situation is inevitably conflict and frustration. Furthermore, managers have found numerous instances where what their systems were measuring was what they could measure, not what they should measure.

A second factor encouraging interest in metrics is a realization that the metrics process defines and communicates what is considered acceptable and unacceptable performance. As a result, metrics have become a major vehicle for directing and shaping the actions of the people on the shop floor. Finally, we are now seeing the emergence of a new metrics approach — an approach that emphasizes the use of metrics as predictive indicators of problems, rather than as outcome or after-the-fact indicators. For these and other reasons, metrics have now become part of the basics of manufacturing excellence. Hence, the reason for this chapter.

The Importance of Metrics (Part 1)

Overview

This focus on metrics is a result of our past experiences. After doing numerous seminars and talking with many managers and executives, we have come to the conclusion that metrics are a major source of confusion. To put it bluntly, we have seen instances of too many measures being used, too few measures being used, the wrong types of metrics being used, and a general lack of understanding of how to use metrics. Such symptoms should be a major concern to any manager because of the critical role played by metrics. Without good metrics, we cannot succeed over the long term. Metrics form the operational link between value (as captured in the mission statement or corporate strategy) and the activities of those people responsible for carrying out activities. To understand the importance of metrics, let's start with a hypothetical story based on a real company.

The Total Quality Management Experience at One Company

There once was a firm that decided to implement a Total Quality Management (TQM) system. Management had invested a great deal of time, effort, money, and manpower on this system. They had hired a consultant. They even went to the extent of creating the position of TQM Champion and put one of their best employees in it. The implementation of the system was kicked off with a great deal of fanfare. Enthusiasm was running high for this new endeavor.

Now, let's move the clock forward some 18 months. The TQM system was not doing that well. The consultant had been fired (or quit). The TQM Champion had asked to be reassigned. Quality levels, which had initially increased, had now fallen back to the old levels. When top management evaluated the progress of their program, they could not help but come to the conclusion that their program was not the success that they had hoped for. They had to find out why. When they talked with various employees, they observed an interesting behavior. To their question regarding the importance of quality, they got the same answers — "important" or "critical." Yet, when they asked these same employees about how they knew they were doing a good job, they got a different answer. Nearly everyone talked about costs and variances.

When management looked into this situation, they saw that performance was measured strictly in terms of cost. Why the focus on cost? The reason was that the measures came from accounting, which had developed a very extensive and well-developed information database. No one had seen the link between strategy and the metrics. What happened was that the people were driven by the cost metrics. Initially, the employees had focused on quality — up to the point that they saw these efforts adversely affect their cost performance. Initially, costs could be expected to go up as employees tried to identify the underlying causes of any quality problem; however, these increases were not acceptable according to the current metrics. What the employees saw was that, while improving quality was good, improving costs was better. The metrics and the corporate intent were at odds. In this company, the two were in conflict, and the metrics won. This result is not unusual. If the strategy and the metrics are in conflict, the metrics will often win.

What Are Metrics?

The first step in effectively and efficiently using metrics is to define what metrics are. Simply stated, a metric is a verifiable measure stated in either quantitative (e.g., 95% inventory accuracy) or qualitative (e.g., as evaluated by our customers, we are providing above-average service) terms. Metrics should be consistent with how the firm delivers value to its customers and should be stated in meaningful terms.

There are several critical features in this definition. First, metrics must be verifiable; that is, anyone should be able to calculate the measure and arrive at the same result if they are given the information used by the metric and the procedure for calculating the metric. Second, metrics are measures. They capture performance in terms of how something is being done relative to a standard. For metrics to make sense, there has to be a standard of comparison. This standard can be based on past corporate experience or on some external standard. For example, we can relate our performance to the best in the corporation or the best in the industry or the best in class (the world-class benchmark). Metrics allow and encourage comparison. This comparison can be between various people or groups or it can be between our performance and this standard. Before leaving this point, it is important to recognize that metrics can be generated at several levels. We can have corporate metrics, such as measures that capture the performance of the overall business (e.g., market share, rate of return, rate of growth). Or, we can have metrics at the product level (e.g., cost per unit, contribution margin per unit, growth in sales). Metrics can also be functionally oriented, in that we can measure the performance of a group such as purchasing. They can be specific to a person or to an activity (e.g., how long it takes to make one unit of output at a specific machine). At each level, we have different requirements. As a result, each level requires its own type of metrics (in other words, one size does not fit all).

Third, metrics should be based on how the firm competes in the marketplace and how it delivers value to its targeted customers. If we are competing on quality, then we should report performance in terms of quality dimensions or activities. As we saw in our example, if we compete on quality but measure cost, then we are doomed to rediscover one of the truisms presented by Oliver Wight (an important consultant and production and inventory control writer of the past): "You get what you inspect, not what you expect."

Finally, metrics have to be expressed in meaningful terms. That is, describing and measuring quality in regard to a stockroom clerk are done differently than for a design engineer. If metrics are to be effective, they must be understood. If they are to be understood, then they must make sense to the person using the metrics.

Why Metrics?

The importance of metrics has been recognized by numerous managers. For example, Tom Malone noted that, "If you don't keep score, you are only practicing." Emery Powell observed that, "A strategy without metrics is just a wish. And metrics that are not aligned with strategic objectives are a waste of time." Finally, some unknown but very wise manager warned, "Be careful what you measure — you might just get it."

Metrics are important because of the functions that they provide, namely:

- *Control:* Metrics enable managers to control and evaluate the performance of people reporting to them. They also enable employees to control their own equipment and their own performance.
- *Reporting:* This is the most commonly identified function of metrics. We use metrics to report performance to ourselves, to our superiors, and to external agencies (e.g., Wall Street, the EPA, or a bank).
- *Communication:* This is a critical but often overlooked function of metrics. We use metrics to tell people both internally and externally what constitutes value and what the key success factors are. As pointed out previously, people may not understand value, but they understand metrics. As a result, value as implemented at the firm should influence the type of metrics developed.
- *Opportunities for improvement:* Metrics identify gaps (between performance and the expectation). Intervention takes place when we have to close undesired gaps. The size of the gap, the nature of the gap (whether it is positive or negative), and the importance of the activity determine the need for management to resolve these gaps.

- *Expectations:* Finally, metrics build expectations both internally (with personnel) and externally (with customers). Metrics help determine what the customer expects. For example, if we say that we deliver by 9:30 a.m. the next day, we have formed both a metric — whether or not we meet the 9:30 a.m. deadline — and an expectation — the customer will be satisfied if the order arrives by 9:30 a.m. We will disappoint otherwise.

Major Lessons Learned

- Metrics are an important but often overlooked aspect of management.
- Metrics are measures that possess certain critical traits.
- Measurement of performance can be done at various levels: corporate, functional, product, activity, or personal.
- Metrics provide certain important functions: communication, control, reporting, identification of opportunities for improvement, and the forming of expectations.

Next, we will focus on the types of metrics, the fit between metrics and strategy, and some additional issues to consider.

The Importance of Metrics (Part 2)

Overview

Our previous discussion on metrics was motivated by our observation that metrics are critical to corporate success but are misunderstood by most managers, and we have seen numerous instances of managers using the wrong type of measures. We will now continue our discussion of metrics by focusing on the different types of metrics and the traits associated with a metric system.

Understanding the Differences Between Metrics

What differentiates an effective metric from an ineffective metric? First, an effective metric should be meaningful to the person using it. An increase in market share does not make much sense to a foreman responsible for managing a filling line. What does make sense to that person is the rate of production, particularly as it applies to actual costs being greater or less than the budgeted costs and the extent to which that manager met due dates and schedules.

Second, effective metrics must be few in number. If we give a manager too many measures, we create confusion. For example, if we were to measure a manager on some 40 different metrics, then we should expect to hear the manager asking us which ones are most important. If we were to reply that every metric is critical, then we would find the manager subsequently focusing on those metrics that he feels he can influence or best manage. As a general rule of thumb, we should try to keep the number of measures to single digits (nine or less).

Third, the metrics should measure activities that the manager can really manage or control. For example, consider a department where the quality of the department is dependent on the quality of the assemblies provided by a supplier. In such a setting, it would not make sense to measure the performance of that department in terms of the quality of the product. That may be something that the manager in that department has little control over.

Fourth, the measures should enable us to differentiate between inherited and created problems. An inherited problem is one that we inherit from another department. A created problem is one that we generate or create by our actions. To understand the differences between these two concepts, consider the following situation. You are managing a department. The system is scheduled using operation's due dates. You receive an order from another department that is three days late. This is an inherited problem. If you process the order normally and it leaves your department still three days late, then you should not be punished for producing a late order. The order was late when it reached you. However, if you were to reschedule the order so that it left your department one day late, instead of three, then you should be rewarded. You have reduced the extent to which the order is late by two days. Similarly, your performance should be negatively evaluated should the order arrive at your department three days late and leave five days late. Your actions have added two days of lateness to the order. This is an example of a created problem.

Fifth, metrics should be timely. That is, they should provide feedback quickly enough so that the people receiving the metrics can do something to correct the problems uncovered. Consider an environment where we receive information about how we did three months ago. This information is not timely. Most people would have difficulty remembering what they did three weeks ago, let alone three months ago. Such measures are punitive, not corrective. They are intended to punish people rather than help them identify problems and correct the underlying root causes.

This last characteristic leads to an interesting discussion. One of the authors was involved with a department that designed and built prototype tooling. It became evident from talking with the various people involved that a major problem facing this department was its inability to generate credible and accurate forecasts. After collecting data about past jobs, the author found that the estimates were, on average, 25% of the actual costs incurred. When the managers involved were presented with this information, they responded that there was not much they could do because of the long lead times. It took some two years for a total project to be completed. By the time the project was done and the actual data collected, it was too late to do anything. It was simply a fact of life, something that everyone had learned to live with. No one was ready to accept the obvious answer — break up the project into smaller, bite-sized portions and measure each smaller activity in a timely fashion.

The sixth and final characteristic to be discussed here is that the metrics should be predictive, rather than output oriented. An output-oriented metric is one that is generated only after the fact. In contrast, a predictive measure is one that we can use to help predict our chances of achieving a certain objective or goal. To understand the difference between these two types of measures, consider the following. Lansing, MI (home of Michigan State University) is located some 90 miles away from Detroit. It is now 12:00 noon and we know that we have to be at the Renaissance Center in downtown Detroit no later than 3:00 p.m. An output measure would involve having someone with a stopwatch waiting for our arrival at the Renaissance Center. If we arrive at or before 3:00 p.m., we would be marked as being on time; however, if we arrive anytime after 3:00 p.m., we would be marked as late. As you can see, with this type of measure we only know how we did when it is too late to do anything — after the activity is done. The problem with this type of metric is that we are doomed to repeat the same mistakes over and over again. There is really no opportunity for us to learn and improve.

In contrast, with a predictive metric we would take a very different approach. We would start by noting our starting time (12:00 noon), the availability of resources (we have a car in good working order which is fully gassed and ready to go), and the average speed per hour (50 mph). With this information, we can predict that our chances of reaching the Renaissance Center before 3:00 p.m. are very, very good. With this type of metric, we can also make adjustments should any of these indicators point to a problem. For example, if we leave later (e.g., 1:15 p.m.), we might have to adjust the average speed per hour to meet our objectives. In some cases, we may have a strong indicator that we will not make our objectives (e.g., we were forced to delay our departure until 1:00 p.m. and, because of construction on the highway, our average speed drops to 30 mph). With two hours to make 90 miles and an average speed of 30 mph, we now know that we are going to be late and can take appropriate corrective action. For example, if we know that someone is waiting to meet with us at the Renaissance Center, then we can try to contact the person to tell him that we are going to be late.

In many firms, the bulk of metrics are output oriented, rather than predictive. As a result, they give managers little ability to predict their chances of meeting their objectives. They only yield information after the fact. For example, a metric such as the percentage of on-time deliveries is an output-oriented metric.

Major Lessons Learned

- Effective metrics are different from ineffective metrics.

■ An effective metric typically has certain characteristics associated with it, the most important of which include the following: they are meaningful, they are few in number, they focus on what the people being evaluated can actually control, they differentiate between inherited and created problems, they are timely, and they are predictive in nature.

■ Many of the measures that most firms currently monitor tend to be output oriented.

Next, we conclude our discussion of metrics by examining the linkage that should exist between metrics and corporate strategy.

The Importance of Metrics (Part 3)

Overview

Previously, we have spent our time understanding what metrics are, why they are important, and the traits associated with any effective set of metrics. This section concludes our discussion of metrics by focusing on the linkages that should exist between the metrics used and corporate strategy. This is a critical linkage because it is here that we see the ultimate power of metrics. If used correctly, metrics become the mechanism used by management to ensure that the actions of people on the shop floor and in the system are consistent with the overall strategic objectives and goals.

Metrics and Corporate Strategy

In any ideal system, there should be no gaps between the intent of the corporate strategy and the metrics that are being used to evaluate and record performance. If we as a firm want to compete on quality and lead time, then we should ensure that we develop measures that track performance in terms of quality and lead times. More importantly, we should ensure that these measures are process rather than output oriented. That is, we want measures that are predictive in nature, rather than being after the fact. We want to give our people the opportunity to identify and correct problems before they become serious.

When dealing with metrics and strategy, however, we should recognize that strategy is nothing more than the specific type of value that the firm intends to deliver to its customers. As such, this value should reflect the manner in which management deals with its customers (identifying critical customers), capabilities (being aware of what the transformation system can do well and what it cannot do well), vision (defining how the firm sees itself in the market), and the environment (taking into consideration the effects of technology, government, and competitors). An effective strategy should try to develop a value statement that brings consistency to these diverse elements. That is, we should develop a value statement that is consistent with our vision and our capabilities, is demanded and prized by our target customers, and differentiates us effectively from our competitors in the marketplace.

If we want to have metrics that are consistent with corporate strategic objectives, the first requirement that must be met is that we have an effective and well-defined strategy. Without such a strategy, it is difficult to set in place the proper type of metrics. Without such a strategy, the metrics that are used tend to reflect primarily internal considerations. We measure what we can, rather than what we should. For example, if we do not know how we compete in the marketplace or what we must do well for the firm to be successful, we tend to measure what we can do well. Assume that a company measures per unit cost or per unit lead time. They improve their performance on these two dimensions by scheduling longer and longer runs. This strategy makes sense to the manufacturing people because it is something that can be readily implemented and done well. Without any guidance from corporate strategy and/or top management, this is what they will focus on. After all, everyone wants to look good when measured.

Gaps can and do arise. When these gaps arise, they indicate the presence of a problem somewhere in the system. When these gaps arise, we must first determine whether we are dealing with a strategic problem or tactical problem.

Gaps as Indicators of Strategic Problems

A gap occurs when the measures that are recorded are not in sync with the desired performance. For example, we think that we are competing on lead time and quality. The measures from the shop floor indicate that our performance on these dimensions is improving — setup times are falling, parts-per-million defects are falling, overall lead time from the moment that the order is received until that order is done and out the door is falling. Yet, there are indications from elsewhere in the firm that these improvements are not enough. Marketing is hearing more and more complaints coming from its customers. Corporate profit and market share are falling. We have a gap. There is a lack of agreement between the metrics being used in operations and the overall performance of the firm. Before turning on manufacturing and hammering them to improve their performance (e.g., "you are not doing enough to reduce lead times" or "your attempts at setup reduction are pathetic — can't you do better?"), we should recognize that the solution to this gap problem may be found at the strategic level.

What we are seeing may be the result of a change in the marketplace. Our customers may have become used to short lead times and improved quality. In the past, these dimensions were considered important premiums, but now they are considered as being normal and simply good enough. Instead, customers we are serving may now want customization, on top of short lead times and improved quality. They may want to have orders shipped to them in smaller, mixed batches. Or, they many want larger batches delivered with longer intervals between them but at significantly lower costs. These changes may be due to shifts in customer expectations. They can reflect competitive actions taken by our competitors (e.g., our largest competitor is now offering to ship mixed loads in smaller quantities and more frequently with customized packaging). They can also reflect a change in technology or governmental action.

When these types of changes take place, the most effective response is a strategic one. Top management must view the inconsistencies between the metrics as symptoms of a strategic mismatch. The solution in this case is to change the strategy or to change the market. In each case, investments are required, as well as lead time. Changing our strategy may mean that we need new and different types of capabilities. To gain these desired types of capabilities may require changing the manufacturing processes or the manufacturing systems. Such investments take time to implement and to mature. Similarly, deciding to change the market or the critical customers that we are targeting requires time and resources to identify new markets or critical customers and to make them aware of our product or capabilities.

Gaps as Indicators of Tactical Problems

In some cases, the metrics may indicate that we are experiencing a tactical problem. The strategy is correct, but we are experiencing a problem in the way in which we are implementing the strategy. When this occurs, managers should recognize that there are four possible sources for this gap: (1) metrics, (2) process, (3) product, or (4) customer.

The first source, metrics, is a simple one. In this case, we are saying that the measures we are using are somehow wrong. That is, they are being applied incorrectly or we are measuring the wrong behavior. For example, a new employee is recording performance by rounding up or rounding down the observations (measures being applied incorrectly). Alternatively, a company could be measuring performance at a non-bottleneck operation, when instead it should be measuring performance at the bottlenecks (measuring the wrong behavior). In this case, the solution to the problem is obvious — change the metric, either how it is applied or what is being measured.

With process problems, the metrics are correct and indicate the presence of problems in the process (this is what we most often think of when dealing with metrics). The proper course of action here is to examine the process and change its management or structure. With product problems, when there is a problem in the functionality offered by the product, we need to examine the product and change it (by adding or dropping features or by changing the services surrounding the product). Finally, with a customer problem, the gaps may indicate that we are selling to the wrong customers. One of the authors saw this problem firsthand in a firm that prided itself on its ability to deliver a quality product. One of its customers had a "variable" definition of acceptable quality. When that customer was under time pressure, it would take and accept anything that the firm could make and deliver. Under these conditions, customer rejects were nonexistent. However, once the crisis was past, the customer would become very picky, with the result that customer rejects often skyrocketed. Management did not know what to do. Ultimately, they decided that the problem was not in the performance of the product, the process, or the metrics; rather, the problem was the customer. This customer created too much variability. The firm recognized that they preferred to deal with customers who wanted the consistent delivery of a quality product, so they dropped this customer.

As can be seen from this brief discussion, the challenge of metrics involves determining whether the metrics gap is strategic or tactical. Both dimensions must always be considered. In most cases, we begin by assuming that the causes of the gaps are tactical; however, over time, persistent metrics gaps strongly point to strategic problems.

Major Lessons Learned

- For metrics to be most effective, they should be consistent with and reflective of corporate strategy.
- Strategy refers to how the firm creates value in the market. As such, it involves four elements: capabilities, vision, the customer, and the environment.
- Gaps can indicate problems that are either strategic or tactical in nature.
- A strategic problem requires that management change the strategy. This change can take the form of a change in the target market or a change in the capabilities or vision of the firm.
- A tactical problem can be due to problems in one of four areas: (1) metrics, (2) process, (3) product, or (4) the customer.

In the short term, most gaps tend to reflect tactical problems. If they are persistent, then they indicate strategic problems.

5 Process Thinking: Key to the Basics

Process Thinking: Understanding the Importance of Processes

Increasingly, we are seeing the emergence of a new approach to the challenge of attaining manufacturing excellence — process thinking. Process thinking is a structured approach that views a firm and its various activities and functions in terms of the underlying processes. With this approach, managers design, document, manage, and change the various processes with the goal of ensuring that these processes make the "desired" results inevitable. For example, when we look at a McDonald's or Burger King restaurant, we see process thinking in action. McDonald's can offer us fast delivery of a consistent quality product because it has standardized both the product and the process (the proof of this can be seen when we ask for a Big Mac made without the sauce and find ourselves waiting forever — why?). Until recently, McDonald's built standard products to stock. When we entered the restaurant and placed our order, we probably looked at the bins containing the burgers to make sure that the items we wanted were in stock. In contrast, Burger King has organized its processes and products around an assemble-to-order approach. It builds components (buns and cooked burgers) to stock. When we enter and give our order, the preparer can assemble the burger our way. To ensure that it is hot, it is microwaved. The reason why McDonald's had difficulty satisfying special orders but Burger King easily could was the structure of their processes.

Process thinking is an approach taking hold at companies such as General Electric, Masco/MascoTech (a Fortune 50 company supplying components to the automotive industry), Lear Company (the world's largest supplier of automotive interiors), Daimler/Chrysler, and Marriott Hotels. It is an approach that is equally applicable to manufacturing processes and service operations. Process thinking represents a new and fundamental change now taking place in manufacturing. In the past, operations managers would focus on activities and resources. Managers would talk about capacity and inventory. They would also talk about activities such as inventory control and inventory management. Now these same managers see that activities and resources, while important, are not as critical as processes. It is processes that determine the specific type of value that a firm and its manufacturing system can offer. To change the type of value offered, the process must be changed. As a result, the task of the modern production and inventory control manager is that of managing processes.

Processes: The Basic Building Blocks of the Operations Management System

Overview

In this and the next three sections, we are going to focus on one of the most important but often overlooked elements of any operations management system — the process. Increasingly, we are beginning to understand that what we do as managers is to design, analyze, document, and change processes. Outcomes such as quality, cost, lead time, and flexibility are not created in isolation; rather, they are the products of processes. To change these outcomes, we must focus our attention on those processes that give rise to them. Put in another way — we are not production and inventory control managers so much as we are process managers.

To understand the overall importance of processes, it is first useful to see this concept in action. One of the world's foremost manufacturers of marine inboard and outboard engines was faced by a problem. Analysis of its market indicated that the market wanted products that were delivered quickly and at a relatively low cost. These requirements made sense given the market and the product. The product, marine engines, was seasonal in nature (if you live in the upper Midwest, then you know that it is difficult to use a boat in the winter). Furthermore, these products are a discretionary luxury. That is, a marine engine is something that you buy when the economy is good and you feel that you can afford it. This means that you often wait until the late winter or early spring to place orders for engines (by this time, you are pretty sure of what the market is going to do).

With these requirements identified, the management decided to examine the processes that they used in building their engines. It very quickly became obvious that the bottleneck process was in the production of the driveshaft housing (the piece connecting the motor to the propeller). When they studied this process, they found that it took some 122 steps (of which only 27 were operations) to start and complete this housing. Furthermore, they noted that this process took some 1500 hours (consider this — one production year of one shift per day, five days a week, equates to 2000 hours). The process itself covered about 20,000 feet and involved 106 people who touched the order in one way or the other. In light of this evidence, management came to one inevitable conclusion — if they wanted to achieve the outcomes of short lead times and low cost, they would have to change the nature of this process. Anything less than changing this process would result in short-term improvements but long-term frustration and poor performance. The key to fundamental and long-term improvement could be found in the process.

What Is a Process?

People deal with processes every day. At a restaurant, a server offers a menu and takes the diner's order as part of the ordering process. The manufacturing process in the kitchen prepares the order. Finally, the order delivery process moves the food from the kitchen to the table. Although processes are central to a company's success, they can generate deep confusion in many managers. Processes involve activities, but they are not the same as activities. Processes draw on structures and resources, but they are not simply structures or collections of resources. A process is a collection of activities that transforms inputs into an output that offers value to the customer. This definition emphasizes five major traits:

1. A process is a collection of activities.
2. A process transforms inputs (physical resources and information) into outputs (goods, services, or information).
3. Structure and capacity determine the resources that the process requires to accomplish its transformation, making issues such as bottlenecks important.
4. A process creates an output that can range from the delivery of an existing product through design and delivery of a new product to the generation of information.
5. Processes are linked to other processes both vertically and horizontally, making them interdependent. This creates a need for careful management of the interfaces between processes.

Processes are critical to every manufacturing company. In fact, every firm could be regarded as being nothing more than the sum of its processes. Further, underlying every firm are at least eight basic processes.

Strategic Management Process

This type of process identifies, formulates, and implements an organization's strategic vision of value and its overall objectives and goals. To generate this output, it must manipulate inputs of information about the organization's unique capabilities, the market's needs, the outputs of rival organizations, and the potential costs and benefits of closing the gap between what customers are getting and what they want.

Innovation Process

This type of process generates new product and process designs. It includes activities to identify a concept or idea, describe that concept, and complete development of the new product or process. Operations managers often group these activities into narrower, supporting processes. The first is the *product design process,* which generates the products sold by the firm. The second is the *process design process,* which designs and implements new processes. Finally, there is the *system design process,* which designs and delivers new systems such as a new computer for an operations management (OM) system.

Customer Service Process

This process determines how the firm or its manufacturing system interfaces with customers, especially external customers. This process covers such activities as the acceptance of a customer order, the tracking of orders, and resolution of customer complaints. Included in it are such processes as *order entry and processing/promising* (responsible for accepting orders and generating order due dates for customers), *forecasting* (predicting what the customers will demand in the future), and *demand management* (matching customer demands with the capacity and capabilities of the OM system so that the customers have credible commitments).

Resource Management Process

This process combines the outputs of the strategic management process (company objectives) and the customer service process (customer orders) to determine the firm's resource needs (capacity, labor, tools, and materials). Included in this process are such supporting processes as *planning* (taking estimates of demands on OM resources and comparing these estimates with current and projected resource levels to identify any unmet needs), *acquisition* (acquiring the resources to meet the shortfalls in capacity identified in the planning process), *scheduling* (setting priorities on orders and assigning resources to orders), and *execution* (implementing the decisions identified in the other processes by releasing orders and working on them to provide the needed goods and services).

Supply-Chain Management Process

This process recognizes that no firm's manufacturing system can ever be completely self-sufficient — nor do they want to be self-sufficient. Every manufacturing system depends on inputs from suppliers through the supply chain, so resource management must extend beyond immediate links to direct suppliers to ultimately reach all sources of the firm's inputs beginning with raw materials. Production managers realize that suppliers provide not only materials, components, and services, but also expertise, market information, and access to new technology. The operations management system of today brings together into a seamless whole the capabilities of its own facilities (internal factory) and those of its suppliers (external factory).

Logistics Management Process

All but the smallest manufacturing systems must solve problems that arise from physical or spatial separation. The firm's logistics management process works to close these geographical gaps by implementing the least-cost combinations of subprocesses such as warehousing and trans-shipping (temporary storage of inputs), transportation links, logistics inventory (inputs in transit through the logistics system), and order delivery/communication (flows of information and goods to meet customer needs).

Performance Measurement

Metrics play a very important role in the manufacturing system. This set of processes determines appropriate measures for determining the structure of processes and for assessing their performance. As noted in our previous discussion on metrics (Chapter 4), the importance of the measurement process can be found in an old saying attributed to Oliver Wight, a well-known manufacturing consultant, who said, "You get what you inspect, not what you expect." Metrics communicate to everyone what it is that they and the processes must do well for the firm to deliver value to its customers. People and systems respond to what they are measured on. Measurement is a communications tool. It tells everyone within the operations management system what it is that they are expected to do well. If performance is measured primarily in terms of costs, people will work at reducing costs. If performance is measured in terms of quality, people will focus on quality.

Other Processes

The preceding seven processes are closely involved in the manufacturing process. There are other processes that indirectly support the operations management system. These processes take place in such departments as accounting (cost accounting, the process of forming and maintaining cost data), finance (the process of evaluating the financial impact of alternative investments), and marketing (the process of formulating and implementing marketing plans).

Major Lessons Learned

In this section, we have identified the critical role played by processes. We have also defined what we mean by the concept of a process. We have identified the critical traits associated with every process. Finally, we have shown how firms can be defined by their processes. Hamburgers that we get from a Wendy's or a McDonald's may seem the same but are differentiated by the processes that produced them. If you do not believe this, then visit various fast-food restaurants and see the differences for yourself. In the next section, we will look at the basic activities that are found in every process.

Identifying the Building Blocks of the Process

Overview

In the preceding discussion, we noted the importance of processes to production and inventory control. In fact, it is very useful for any manager to become a "process thinker." Any manager facing a problem should immediately try to identify the processes associated with that problem, document them, study them, and then try to determine how to best improve them. At times, this way of thinking requires that we simply refine the current process. That is, we take what is being done in the current process and merely make it better. We identify the sources of waste or redundancies and eliminate them. We take areas of inefficiencies and make them more efficient. In other cases, we have to redefine the process. With this approach, we recognize that any changes that we introduce into the existing process are merely Band-Aids. The process is fundamentally wrong. There is nothing that we can do to improve it. Faced with such a system, we have to go back to the drawing board. Starting with a clean sheet of paper, we redraft the process from scratch.

Regardless of the approach that we choose, however, we must be able to document the existing process or processes. At this point, we have a challenge. We want to avoid using words to describe the elements of a process. Words, when everything is said and done, are often imprecise, ambiguous, and misleading. They are open to interpretation, and whenever there is an opportunity for interpretation there is also a risk of misunderstanding. What would be more useful would be symbols that help categorize the various activities taking place in a process.

Symbols offer numerous important advantages. Because they are graphic in nature, they are concise. We can replace numerous words with a single picture (remember the saying about a picture being worth a thousand words). They can be used to develop maps of processes that we can study when performing process flow analyses. Finally, they lend themselves easily to computerization and the use of computer software.

Before introducing the symbols, we must keep in mind that our goal is to develop a comprehensive, mutually exclusive, and concise classification system. That is, we want to be able to identify and categorize every activity that takes place in a process (the criterion of being comprehensive). We also want to have a system in which an activity assigned to one category or symbol cannot be assigned to another category or symbol (the criterion of mutual exclusivity). Finally, we want to use the fewest number of symbols. With fewer symbols, we have less opportunity for misunderstanding and a system that is easier to learn and easier to teach.

With these criteria in mind, we turn to a set of categories that were initially developed by industrial engineers in the 1950s. Please remember that this is just one approach — there are others. However, we have used this system and we know that it works. With it, we can effectively and efficiently categorize every activity that takes place in a process.

Categorization of Activities within the Process

This categorization system uses five categories to capture all of the activities that take place within a process:

1. Operation
2. Transportation
3. Inspection
4. Delay
5. Storage

Operation

An operation is any activity that causes a change or transformation to an input. An operation occurs whenever:

- An activity intentionally changes one or more of the major traits of an input. For example, workers on the line at an assembly plant who add trim to a van's exterior are involved in an operation. They have changed the trim and the van by merging the two and making the end product closer to the product desired by the customer.
- Information is communicated. An operation takes place when an employee is told how many vans to build or the options required per van.
- Planning or calculations take place. An operation might determine how many vans are to be built over the next month.

For the most part, operations are the major source of value in most production and inventory control systems.

Transportation

Transportation is any activity that moves an object from one place to another. A transportation activity changes only the location of a part, without transforming its characteristics.

Inspection

An inspection checks or verifies the results of another activity. An inspection might examine a part to determine if it is the correct item or to compare it against a standard. Frequently, the standards used within this activity take one of the following forms:

- *Internal standards:* An inspection might compare actual production times against standard times (the internal standard). Alternatively, it might compare a van assembly against the standards set down in the engineering drawings.
- *External standards:* These are standards that come from outside of our firm. In some cases, these standards might come from our customers. They might also come from the government or official organizations (e.g., Underwriter's Laboratories or the International Standards Organization, ISO).
- *Benchmarking standards:* Benchmarking is a technique by which managers assess their company's performance against either that of other companies or the current best practices.

With an inspection, we are deciding whether a good or service passing through an activity is acceptable or not. We are also deciding whether the activities involved in the process are proceeding in an acceptable manner.

Delay

Delays result from interference. For example, suppose you arrive at the front desk of a hotel after a long and tiring flight, and you want to check in. Before you can do so, however, the clerk at the front desk must finish registering the people ahead of you. You are experiencing a delay. Delays can occur before work waiting in line. They can also take place when orders have to wait because of missing information or because other parts necessary to complete the order have not arrived on schedule. Delays can also occur because of such unanticipated problems as machine breakdowns or accidents. In all of these cases, a delay is a temporary stoppage. It describes a good or service momentarily at rest.

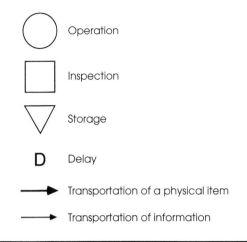

FIGURE 7. Categories of Activities

Storage

Storage is any activity where items are stored under control. To gain access to this activity requires appropriate authorization. For example, whenever we put money in a bank or credit union, we are putting our money into storage. It is a storage because if we want to get our money out we need to first give the teller at the bank a withdrawal form. This withdrawal form can be a withdrawal slip, a check, or an ATM (automated teller machine) card. Without the appropriate authorization, the bank will not release any money to us. In operations management, storage activities can take many forms. They can be stockrooms located on the shop floor, they can be warehouses, or they can be holding/receiving areas.

Storages resemble delays, but they differ on the dimension of control. A storage must include the *formal* control of an item; delays have no such requirements. A door panel sitting in a stockroom is a storage (because access to it is strictly controlled and release can only be made upon receipt of proper authorization). When that panel reaches the floor but waits to be installed, it becomes a delay.

One way to decide which category to select for describing a specific activity is to use the word associations that are shown in Figure 7 and below.

Operation	Produces, does, accomplishes, makes, uses
Transportation	Moves, changes location
Inspection	Verifies, checks, makes sure, measures
Delay	Interferes, temporarily stops
Storage	Keeps, safeguards, protects

Major Lessons Learned

We have learned in this discussion that there is an advantage to describing activities using a system of categories. With a system that is graphic in nature, comprehensive, mutually exclusive, and concise, we can begin the task of bringing processes under control. All the activities that take place in any process can be captured using the five categories of operation, inspection, transportation, delay, and storage.

Structure of Processes

Within any OM system, there are many processes. Some processes consist of subprocesses. These processes, in turn, combine to form still larger and broader processes. Ultimately, the result is a mutually reinforced network of processes that we know as the production system. Regardless of whether we are dealing with the very focused, smaller subprocesses or the larger network of processes, analysis and description of the process remain the same. Process structure deals with the organization of inputs, activities, and outputs of a process. Operations managers define this structure by ordering activities, positioning them, and linking them.

The Ordering of Activities

Activities rarely occur randomly. Some must occur before others for the process to flow smoothly. The requirements of the process determine this precedence or sequence of activities. For example, a cook must make pancake batter before cooking the pancakes on the griddle. A process that can operate only in a specific sequence requires *strict precedence;* the cook cannot possibly make pancakes before preparing the batter. Other activities can occur in more than one sequence, allowing *nonstrict precedence.* For example, a computer assembly process could insert microchips in several different sequences. Whether a process requires strict or nonstrict precedence of activities, operations managers must order activities carefully to achieve effective performance measured by lead time, cost, quality, and flexibility.

Positioning of Activities

Ordering determines the sequence of activities over an entire process, while positioning organizes activities in relation to each other. A process can position most activities either *sequentially* or in *parallel.* Sequential positioning places activities one after another, as in Figure 8. A pancake cook might position activities sequentially by opening the recipe book, gathering the ingredients, mixing them, cooking the batter on the griddle, and delivering the pancakes to

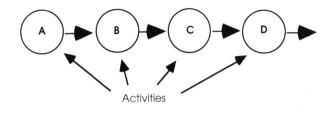

Activities

FIGURE 8. Positioning of Activities Sequentially

the waiting diner. Sequential positioning results in a total process lead time equal to the sum of the lead times of the individual activities.

Sequential positioning also limits the responsiveness of a process when it must cope with changes. Any change must often be introduced at the initial activity and carried in sequence through all of the other activities. In addition, when a problem caused early in the process becomes evident only later, correcting it requires returning to the activity where the problem was created and moving the solution back through each later step.

In contrast, parallel positioning organizes activities to occur simultaneously as much as possible, as illustrated in Figure 9. For example, the pancake cook might organize the cooking and consumption activities in parallel so diners can eat one batch while the next is cooking. Parallel positioning tends to reduce lead times, as it determines total process lead time according to the sequence of activities with the longest lead times (often referred to as bottlenecks).

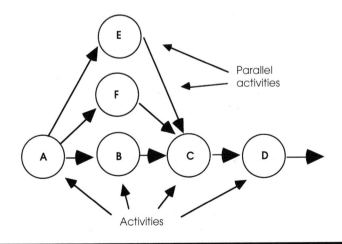

Activities

FIGURE 9. Positioning of Activities in Parallel

Linking of Activities

Links result from the relationships created as activities are positioned. A *spatial link* represents the distance (measured in feet, meters, inches, or similar units) between two activities. Locating two related activities closer to one another should reduce the time required to move components, tools, or other resources between them.

A *physical link* represents a tangible connection between related activities. For example, an automobile assembly line physically links activities by placing workstations in a specific configuration and moving cars past them in a set order. Physical links often reduce lead times. Also, lead times show less variability (the difference between actual lead times and their means) because the physical connections allow little or no deviation from the standard sequence. After the initial expense of building a physically linked production line (a fixed cost), such linkage can generate lower variable costs than less tightly linked processes. It also reduces material handling (as the physical link handles the products). Finally, inventory is often limited to what can be stored within the physical links. Physical links also limit flexibility, however. If some change requires activities to occur in a different sequence, one must often change physical links, tearing up existing production lines and building new ones. This consumes considerable time and money.

The performance of any process depends substantially on the resource flows defined by links between activities and their sequence. Inputs also have significant effects. The capacity of the process joins with these elements to determine the performance of a process (see Figure 10).

Major Lessons Learned

In this section, we explored the nature of processes in greater detail. Specifically, we have shown that all processes consist of subprocesses; all processes feed into larger processes. Further, we have shown that process performance is influenced by the type of activities previously discussed and the issues of structure. Processes can be structured to be sequential or parallel. The type of structure has major implications for such dimensions of performance as lead time, cost, and impact of bottlenecks.

Improving Processes: The First Step

Overview

Up to this point, we have focused our attention on illustrating the importance of processes, defining what a process is, and identifying the major categories of activities found in every process. All of these activities, while important, are

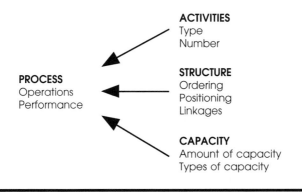

FIGURE 10. Factors Affecting a Process

not enough in themselves. They are simply the first steps that we take to improve the performance of the processes. However, before we can describe the steps that we can take to improve the process, we must first discuss the need for management to close the gap between the four processes found in every firm.

Four Types of Processes Found in Every Firm

In every firm, every process exists in four different forms:

1. The process as formally documented
2. The process as people think it exists
3. The process as it actually exists
4. The process as it should exist

Between these four forms, there exist gaps. One of the objectives of process thinking is eliminating these gaps between the various forms of a process.

The Process as Formally Documented

Most firms and OM systems attempt to document their processes (especially the critical ones). This is done for a number of reasons. For example, customers may require documentation of the processes. Also, when a company is trying to achieve certification (as in the case of ISO/QS 9000 quality certification), one of the requirements for certification is to have formal documentation of their processes. In other cases, a company may document their processes so that they can be taught to others — new employees, customers, or suppliers. Unless the documentation is continuously maintained, however, documented processes tend to lose their accuracy over time.

This loss in accuracy occurs for several reasons. The process can be changed as a result of a change in technology or a change in the design of the product. It can also change in response to problems encountered within the process. For example, suppose we are building a monitor and have experienced quality problems with parts being delivered by certain suppliers. In the short term, our response is to insert a quality inspection station into the process so that all the parts coming into the process are first checked (note that in the long term we may not need this inspection if we work on eliminating the quality problems taking place at the supplier's location). Finally, the process may change because of informal improvements introduced by people who work with the process. Over time, these people may discover and implement opportunities for improvement without telling the people who document the process. As a result of these factors, a gap develops between the process as documented and the process as it exists.

The Process as People Think It Exists

The process that actually exists is often different from the process that people think exists. Everyone knows, for example, that master production scheduling involves production, marketing, scheduling, purchasing, and finance/accounting (with the major activities being carried out by scheduling). If we were to ask any of the groups not directly involved in the actual process of generating the master schedule how it is done, they would probably tell us what they think happens. For the most part, though, they would be wrong. The reason is that most see only a part of the entire process — the part that they are involved in. They know this one part well, but they must make assumptions about what takes place in the other parts of the process. These assumptions are often based on observations that people make about the other areas. These observations, in turn, are often drawn from their experiences with these other elements.

The Process as It Actually Exists

With this form of the process, we are interested in the process as it actually exists. Often, we describe this process as the base or "as is" process. One of our objectives should be to reduce the gaps between the first three forms of the process.

The Process as It Should Exist

The difference between this form of the process and the preceding one is value. With this form, we have identified and changed all of those activities in the process that are either inefficient or ineffective (i.e., do not add value). It is our goal to develop processes that are inherently able to deliver the type of value expected or prized by the target customers.

It is critical to understand these four forms of a process because of the problems they cause. Because of the gaps or differences between the first three forms, we find ourselves faced with confusion, frustration, and inefficiencies. People become confused when what they think happens is not what actually happens. These differences in perception often lead to frustration and increased inefficiencies (as everyone acts on their belief of what is happening rather than what is actually happening). One of the first things that managers must do is to close the gaps between these first three processes. This is supposedly one of the major reasons why firms undertake ISO/QS 9000 certification — to ensure that the documented processes are identical to the actual processes in use. However, closing the gaps between the first three processes, while important, is only a starting point. The next and more critical step is that of closing the gap between the third and fourth form of the process. This is the task that we will discuss next.

Major Lessons Learned

We have shown that in nearly every firm every process exists in four different forms: (1) the documented process, (2) the process that we think exists, (3) the process that actually exists, and (4) the process that should exist. The presence of these gaps creates many hidden costs for the production and inventory control manager and for the overall system as a whole. Our task is to identify and close these gaps. The first step in closing these gaps is to identify and document each of the forms that a process can take.

Incorporating Value into Process Analysis

Overview

Up to this point, we have focused on describing the processes as they exist and on closing the gaps between the first three of the four possible forms of processes that can exist. However, we have not yet addressed the last gap — the gap between the process as it currently exists and the process as it should exist. To close this gap, we must introduce the value concept into our analysis of process. By introducing value, we examine the extent to which the process is able to meet or exceed the requirements, needs, and expectations of the customer. It is here that production and inventory control and production can create a strategic and market advantage for the firm. Introducing value into process flow analysis is largely a straightforward task consisting of the following steps:

1. Identify the critical customer/product traits.
2. Determine what constitutes value for the critical customers.
3. Identify metrics consistent with these specific traits of value.
4. Identify the critical processes (including bounding and setting base conditions).
5. Document (and validate) the process as it currently exists.
6. Assess the performance of the current system in light of the customer-based metrics.
7. Modify the process appropriately.

Identify the Critical Customer/Product Traits

Before we can effectively undertake a process analysis, we must first understand both the customers and the products that we provide. Processes have many customers. Some customers are internal — for example, in the assembly line setting, the person stocking the parts bin has an internal customer in the form of the people who take parts from the bin. There are also intermediary customers — for example, the shipper who moves the cars from the plant to the various dealerships. Finally, there are external customers, such as the dealerships and the people who take delivery of the van. Each customer places its own demands on the process. As a result, the first step in the value-driven approach is to identify the key or critical customers. A critical customer can take many forms: a customer that we would like to sell more to in the future, the customer that generates the largest amount of revenue (either in aggregate or per transaction), or the person who has the greatest impact on how much we sell.

In identifying the critical customers, we determine which customers we must do an excellent job of satisfying (by either meeting or exceeding expectations and demands). By targeting critical customers, we focus our attention. When we evaluate the value capabilities of our processes, we do so from the perspective of these critical customers. We invest time and effort in learning as much as we can about these customers. The other customers are still important; however, our focus is to keep our critical customers happy.

Similarly, we understand the critical traits associated with the product that we provide. These traits may affect how we view and analyze the process. They also shape our expectations of the process and help flag potential problems in the process. Typically, critical product traits include the extent to which the product is perishable, the extent to which the demand for the product is seasonal or constant, and the potential for obsolescence. Perishability refers to the extent to which a product can be stored for any period of time. This trait applies to both inputs and outputs. Perishability of inputs means that the process must transform inputs quickly (otherwise they will no longer be usable). Perishability of outputs means that the time between the creation of the product and its consumption must be kept to a minimum.

Determine What Constitutes Value for the Critical Customers

Having identified the critical customers, we next identify their expectations of the process. We can identify these expectations in many ways. We can observe the critical customers close up and see how they interact with the processes and how they use the products produced by the processes. We can use interviews and surveys. Finally, we can base these expectations on benchmarking.

Identify Metrics Consistent with These Specific Traits of Value

We convert these customer demands into specific metrics. For example, if we are a mail-order company such as L.L. Bean, we know that our customers want speed; however, this notion of value is vague. If we decide, though, to measure the percentage of telephone calls answered by the second ring, we have defined a measure that can be tracked. We have also identified a specific expectation that the customer has of our process. To emphasize a point previously made, metrics make value real, not theoretical. Finally, it is important that these metrics be established before examining the process.

Identify the Critical Processes (Including Bounding and Setting Base Conditions)

This repeats a point previously made. Because we cannot feasibly study all processes under all possible conditions, we must focus our attention on certain processes operating under certain conditions (e.g., deciding whether or not to include the processing of rework). These critical processes can be those with which the customer has direct contact or ones that affect what the customer receives (e.g., bottleneck processes). In many cases, these processes involve activities such as order fulfillment, product design, manufacturing, or logistics.

Document (and Validate) the Process as It Currently Exists

Having selected the customer, we can then document the processes.

Assess the Performance of the Current System in Light of the Customer-Based Metrics

We can assess the process to determine the extent to which its outcomes are consistent with the demands and expectations of the targeted customers. We can also assess the extent to which each activity contributes to value by using a categorization scheme consisting of the following:

1. *Value-adding (V):* The activity contributes to value as defined by the customer. Operations frequently fall into this category.
2. *Non-value-adding but necessary (N):* The activity does not contribute to value but is needed by the process or by other activities within the process. Setups and the occasional inspection fall into this category.
3. *Waste (W):* The activity consumes resources but does not increase value. Inspections, moves, storages, and delays often fall into this category.
4. *Unknown (?):* The nature of the activity has yet to be determined.

The assignment of values to the activities is often done using the "five whys" technique, which requires that we ask "why" five times. The first time that we ask "why," we will either get what the respondent thinks we want to hear or a general statement. By the time we ask "why" for the fifth and final time, we will finally arrive at the real nature of the category (i.e., the extent to which it contributes to value).

Once these activities have been categorized, we have some indication of what we should do with the activities in terms of disposition. Ideally, we should eliminate activities that result in waste; we should reduce the time and resources needed by non-value-adding but necessary activities by either rethinking or combining. The value-adding activities should be either kept or rethought if there is an opportunity for improvement. Finally, we should categorize all unknown activities into one of the preceding three values (V, N, W).

Modify the Process Appropriately

The final stage is to change the process by either modifying it or replacing it through a process such as Business Process Reengineering. One comment about this final stage should be made. In practice, this process results in observations about other areas for improvement that lie outside of the scope of the current project. These observations are opportunities for improvement that can be acted upon in subsequent projects focusing on improving the effectiveness and efficiency of our processes.

Major Lessons Learned

We have begun to examine the process by which we close the gap between the process that actually exists and the process that should exist. To close this gap, we have introduced value into the process flow analysis. We also described a process aimed at helping managers close this gap. At the heart of this process is the assignment of every activity to one of four value-based categories: (1)

value adding, (2) necessary but not value-adding, (3) waste, and (4) unknown. With this assignment, we are now ready to close the gap between the process that actually exists and the process that should exist.

Process Analysis: Closing the Value Gap

Overview

To close the gap between the process that exists and the process that *should* exist, we must deal with the twin issues of disposition and position.

Disposition Tactics

These tactics evaluate each activity within the process and dispose of the activity in one of four ways:

1. *Keep.* This decision leaves intact any current activity that the analysis designates as both efficient and effective, so no improvements seem appropriate.
2. *Combine.* This decision joins an activity with others that do the same or similar things to improve the efficiency of the process.
3. *Rethink.* This decision reevaluates an inefficient but necessary activity to maintain any contribution to value while enhancing process efficiency.
4. *Eliminate.* With this option, we are faced with an activity that consumes resources without generating any offsetting benefit or value. The activity is then eliminated. This results in a simpler process (through the elimination of steps), shorter lead time (again due to the elimination of steps), and released capacity (this last effect is important only if the capacity released by the eliminated activity can be used elsewhere within the firm).

With the disposition tactics, we are identifying what we should try to do with each activity that takes place in the process. With *keep,* there is nothing that we are going to do with a given step (for the time being; remember that this disposition can and will change in the future, as processes, like people and value, are inherently dynamic). With *eliminate,* the option open to us is fairly straightforward — we are going to improve the process by eliminating that activity. *Rethinking* is the most challenging of these tactics in that it forces us to go back to the drawing board, so to speak. We have a well-defined outcome, but the manner in which we have previously achieved this outcome is no longer appropriate. We must come up with a new method for achieving this outcome — a method that achieves the same outcome with less time, resources, or steps.

In general, we can evaluate the performance or effectiveness of any process using the following set of broad metrics:

1. Number of steps (broken down by category)
2. Distance covered (both vertically and horizontally)
3. Time required (minimum, maximum, average, variance)
4. Value orientation of the activities (value-adding or not)
5. Number of departmental boundaries crossed
6. Number of departments involved in each activity
7. Number of people who touch or come into contact with the order or process

Each of these measures relates to the four components of value: lead time, cost, quality, and flexibility. For example, a longer distance increases the expected lead time and reduces the flexibility of the process.

Position Tactics

In every process, one set of activities determines the overall process lead time. Production and inventory control managers typically refer to this important sequence of activities as the critical path of the process. Adding activities to the critical path increases the overall lead time of the process; removing activities from the critical path reduces overall lead time. The change initiative can position each activity either on the critical path or parallel to the critical path. As discussed earlier, organizing activities sequentially produces a longer process lead time than organizing them in parallel.

A Caution Regarding Process Change

Often, we assume that difficulties observed in the process (e.g., long and variable lead times, unpredictable completion rates, poor quality, and customer dissatisfaction) are due to problems within the process. This is not always the case. Difficulties observed within a process can reflect the impact of one of four factors (see Figure 11):

1. *Problems in the process.* This is the basic or default factor and is often the starting point or assumption; however, it may not be the correct starting point.
2. *Problems with customer mix.* In some cases, the problems observed within the process reflect the inability of the process to satisfy differing and conflicting demands coming from the various customers that it is trying to serve. For example, suppose we serve two basic customer segments. The first gives us orders with plenty of lead time. What they want is low cost and timely delivery (e.g., delivery schedules are always

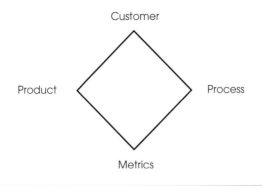

FIGURE 11. Four Root Causes of Poor Process Performance

met). The second segment consists of customers who have emergency orders, the arrival of which is difficult to predict. When the orders do arrive in the system, the customers want immediate service (cost is no object). One system cannot serve both customers. The schedules that we generate to serve the first set of customers would always be interrupted by the second set of customers. When faced by this problem, we must decide whether we will focus on only one customer segment or whether we will develop two separate processes (one dedicated to serving each type of customer).

3. *Problems with product and product design.* In this case, the problem lies with the product. Fundamentally, the product itself does not meet the needs of the customer. No amount of tinkering with the process can correct this basic, underlying problem.

4. *Problems with metrics.* With this option, the problem is not in the process but in the measures that we use to manage and control this process. For example, our customers want speed and on-time delivery; however, we measure the process and the personnel who work within it on the basis of cost. As a result, we will not get speed and on-time delivery but rather low-cost production. The answer is to change the metrics that are used.

Major Lessons Learned

We have examined how the tactics of disposition and position can be used to close this last gap. We also noted that, at times, what we see happening in the process may not be due to problems with the process but instead may reflect problems originating outside of the process. It is important that we be aware of these factors, as they are the real root causes that must be addressed before we can expect to see true improvements in the process.

Poka-Yoking Our Activities

Overview

Many of the ideas discussed in this book have typically originated with either an e-mail message or a telephone call. In some cases, though, they have come up because of a conversation that has taken place in our office, which is how the idea for this section emerged. Recently, a manager working with a leading-edge manufacturer of flight simulation systems was talking with us about some problems the company was experiencing. During the conversation, the manager asked if we could recommend a rule regarding lot sizing. This seemed to be an unusual request. After all, anyone who has had any exposure to manufacturing and production and inventory control has been taught the mantra that "we shall not lot size unless we must."

Before answering this question, we asked the manager to tell us about the reason behind this request. As the manager explained it, inventory had increased rapidly within the company. Over the last two years, the rate of growth in inventory was between two and four times that of the growth in sales. One of the areas identified as being at the root of this problem was the stockroom. After reviewing procedures within the stockroom with the stockroom personnel, an area targeted for further study was setting order quantities. Apparently, both production and purchasing ordered quantities using the rule of NRN (nice round numbers) or placed an order for the largest quantity necessary to get the largest price discount. Given that the firm was in a high-technology environment, such policies created two outcomes. The first was the ordering of a lot of inventory; the second was a high level of obsolescence. Both of these outcomes were highly undesirable.

In the process of talking this request over with the manager, we asked the manager a simple but critical question: "What is the desired outcome of the stockroom?" The manager seemed to be surprised and taken aback by this question. His first response was that he viewed the stockroom as a place where inventory was stored. This implied a passive role for the stockroom — not a very attractive state. After further discussion, the manager identified three major desired outcomes for the stockroom:

1. To ensure access to inventory that was listed in the computer system as being available. This meant that there had to be a close agreement between the inventory records and the physical stocks. If the computer system stated that there were five units of a part in stock, then the people on the floor had a right to expect that there would be five units of that part in stock in the stockroom. That required that the stockroom personnel recognize the importance of inventory accuracy and stockroom control.

2. To provide timely access to the requested items in inventory when requested. Accuracy, by itself, was not enough. When a person came with a properly authorized requisition, then that person should expect to be given the requested parts within a short time period. After further discussion, it was decided that five (5) minutes was a reasonable time interval from the time that the requisition was submitted until the time that the parts were delivered to the waiting person.

3. To act as asset managers. This meant that the stockroom personnel had to take it upon themselves to manage the stocks with the goal of identifying those items in stock that were obsolete or slow movers.

After having identified these three desired outcomes, the manager was next asked to identify those processes that were in place to ensure that these outcomes inevitably occurred. Again, the manager had to think for a moment. His answer was disturbing — he was not sure if there were any processes in place that ensured that these desired outcomes became inevitable. Finally, we asked the manager, "What measures do you have in place that track the performance of the stockroom on these desired outcomes?" Again, the manager had no answer. The only measure used was inventory accuracy (which was done on a periodic basis). Inventory accuracy was measured primarily in dollar terms, not unit terms (not the best way of measuring inventory accuracy).

When this meeting ended, the manager walked away with a new appreciation of the problem before him. The task was not one of finding the best lot-sizing rule. Rather, the task before the manager was to develop a system in which the desired outcomes for an activity were identified in advance, the appropriate processes necessary to achieve these outcomes were in place, and the appropriate measures assessing the success of attaining these outcomes were identified. The nature of the task had changed dramatically.

The Importance of Knowing the Desired Outcomes

In this situation, we see three major tasks facing any manager: (1) identify the desired outcomes of any process or activity, (2) determine that the processes necessary to ensure attainment of these outcomes are in place, and (3) have appropriate measures of success in place. Each of these three tasks is critical for success. The first is important because few people ever clearly articulate (describe) what they expect from a given activity or area such as a stockroom or dispatching. In most cases, people have their own expectations of these activities or areas — expectations that they think are shared by others around them (but which are not). By articulating these desired outcomes, we force managers and others involved with the activity or area to describe clearly and concisely what they think the activity should do. Often, what managers find is that the activity that they are studying is doing things other than what it should be doing. This occurs because no one ever told the people involved what they should do. As a result, the personnel respond to the demands placed on them by their assumed users.

The second task really comes to us from Just-in-Time (JIT) manufacturing. In JIT systems, we are introduced to the concept of poka-yoke (foolproofing). This concept can also be defined as making the desired result inevitable. This is an important concept because it forces us to think of the desired outcomes not as drivers but as residuals. They are the products of processes. The challenge facing management is that of ensuring that the appropriate processes are in place. In the absence of these processes, we find ourselves relying on people. Whenever we rely on people, not processes, then we can expect to see the emergence of an informal system. When this system emerges, then we, as managers, have lost our ability to formally control the activity in question.

The third task, that of putting in place the appropriate metrics, is also critical because we need the information provided by metrics. We need to know if we are achieving the desired outcomes, and we need to know where to look if we are having problems in achieving these desired outcomes. We need metrics to establish expectations of the desired outcomes, both internally (within the activity) and externally (within management and within the users of the activity). We also need metrics to communicate objectives and facilitate feedback.

When these three elements are put together, we find that we have introduced a system for enhancing the performance of activities. With this system, we can focus on addressing problems rather than dealing continuously with symptoms.

Major Lessons Learned

- Be clear on the expectations of any activity or area. You should be able to articulate the desired outcomes for that activity.
- Ensure that the proper processes are in place to make those desired outcomes inevitable (the poka-yoke concept). If they are not, then your challenge as a manager is to make sure that these processes are identified, developed, and implemented.
- Provide feedback on the extent to which these desired outcomes are attained through the use of appropriate metrics.

These points are simple enough, but they are the stuff of which basics are built.

6 | Capacity: You Can't Build It if You Don't Have the Capacity

Capacity: The Resource at the Heart of Manufacturing Excellence

apacity is central to nearly everything that is done in manufacturing. Whenever a decision is made to change a process or to reschedule work, there is also an accompanying implication for capacity. We define or evaluate feasibility of a schedule or a proposed change in terms of capacity. In many ways, the development of manufacturing excellence is really the story of the development and maintenance of effective capacity planning and management.

Effective capacity planning and management are not easy to develop and implement. In part, this is because the concept of capacity is not an easy one to define and understand. There is not just one type of capacity; rather, there are multiple different types of capacity that we must be aware of. In addition, there is the issue of assessing capacity in isolation and within the context of a process. Finally, there is the interrelationship between capacity planning management and production scheduling.

Because of the important role played by capacity, an understanding of the basics of capacity is essential to achieving manufacturing excellence.

Capacity: Understanding the Basics

You would be amazed at how many companies have problems scheduling their manufacturing operations. They seem to be stuck in a mindset of believing that if they have idle equipment or labor, then they can accept more work. This additional work is then released to the shop, but, instead of output increasing, it either remains constant or falls. Frustration increases. In these plants, the major problem is that the management either does not understand capacity or has an incomplete or inaccurate understanding of it. As a result, if we are to correct this problem, it is important that we understand the concept of capacity and why it is so important to production and inventory control practice.

Let us begin with a simple observation. To build a part, we need information (routings, product structure, quantities, timings, and such), material, and capacity. Of these various elements, capacity is often the most critical. In many instances, we reduce the various activities found in most production and inventory control systems to being no more than variations on capacity planning, management, and control. Scheduling, for example, can be regarded as being a capacity management issue. If we manage and measure capacity well, then we can enhance the ability of our shop to build orders in the quantity required, at the time demanded, and to the levels of quality desired by our customers at acceptable levels of cost.

Capacity, however, is a manufacturing contradiction. Every production and inventory control person has had contact with capacity; yet, few are able to understand the various elements of capacity that make it so inherently complex and difficult to comprehend. To understand capacity, we must first begin by defining capacity and then look at capacity at a micro (work center) level before exploring capacity at the shop level.

Capacity exists on two different levels: strategic and tactical. At the strategic level, capacity is capability or what the shop or work center can and cannot do. At the tactical level (which is the focus of this section), capacity can be defined as being the level of output per unit of time. This definition identifies several important traits of capacity. First, capacity is always measured in terms of time. For example, suppose we tell you that there are two work centers each producing 5000 units. This information is either meaningless or potentially confusing. Why? What if the first work center could produce 5000 units in 24 hours, while the second required 120 hours to produce this same number. Now we see that the work centers are different, even though they produce the same number of units.

Time is also important when discussing capacity because capacity is finite with respect to time and cannot be stored. This last fact is something that is very familiar to every restaurant, bank, or similar service operation. When seats in a restaurant are available and empty at 10:00 in the morning, that unused capacity cannot be stored until it would be needed for peak traffic. Capacity that is not used during a period of time is gone.

We can look at inventory as being a form of stored capacity. When we build inventory, we convert the capacity into product which can then be used to meet demand during peak periods.

When dealing with capacity, we must consider two specific types. The first is capacity that is based on resources. A work center can consist of a machine that is available 24 hours a day, 7 days a week. It can also consist of a worker who operates the equipment. Finally, the work center can also consist of the various tooling that is needed to allow the equipment to process the various orders that pass through it. Each of these components has varying levels of availability. Each also defines a type of capacity. For the work center, the resource capacity is often defined by the limiting resource. For example, if the machine is available 24 hours a day but is operated by an operator for only one 8-hour shift, then the capacity of the work center is based on the operator resource capacity.

The second type of capacity is defined in terms of output, or the number of pieces produced by the work center. This definition of capacity is more frequently used in planning. The amount that is produced is often difficult to calculate because it reflects the impact of product mix (the greater the variety of products produced, the less output because of the need to set up when going from one product to another), the nature of the products being built (well-established products should consume less capacity than either prototypes or products that are relatively new), operator experience and capabilities, amount of work from preceding work centers (one work center could be idle because the preceding work centers have not produced enough material for it to process), maintenance of the equipment (a lack could lead to breakdowns), and problems with material and/or tooling. These factors account for a work center being able to produce more in one period than in another. They also contribute to the volatility frequently associated with capacity.

This discussion forces us to deal with the issues of how we measure capacity. There are three measures of capacity, the first of which is *maximum capacity*. This is the maximum amount that a work center can produce. It defines the upper limit of output. We previously noted that capacity is finite; that is, we have a set amount of capacity (the upper limit is the maximum capacity). We use this capacity to satisfy various needs — processing orders, maintenance, setups, or idle time. Because capacity is finite, to increase one component of capacity usage means decreasing the usage somewhere else. If we increase setups, then we can expect one or more of the other components to fall (e.g., capacity used for processing, idle capacity, or capacity set aside for maintenance).

Next, we have *effective capacity*. This is the amount of capacity that we plan for. Typically, effective capacity is expressed as a percentage of maximum capacity. For example, we might state effective capacity as being 75% of maximum. This leads to an interesting and important question — why would we plan for less than 100% capacity utilization? The answer is that we might want to plan for unused capacity so that we can improve the performance of our equipment (running equipment at 100% might create operating problems and unplanned breakdowns), provide safety capacity available for unexpected orders, or allow time for preventive maintenance.

The third measure of capacity is *demonstrated capacity*. This is the level of output that we actually get. While maximum and effective capacity are used for planning (that is, they are before the fact), demonstrated capacity is observed after implementation of the plan (after the fact).

Why collect these three measures of capacity? The answer is that they tell us a lot about what is happening or the types of problems that we are facing. For example, if demonstrated capacity is less than effective capacity, then we know that we are not producing as much as we planned for. We do not have a problem with the availability of capacity but rather with how it is used. For example, the problem could be the job mix or the rate at which work is coming to us from the preceding work centers. However, if demonstrated capacity exceeds effective capacity planning and starts to approach maximum capacity, we begin to see that we are reaching a capacity problem (we might not have enough capacity). We must now try to find out why. At this point, we are faced by an important consideration — how to measure capacity, which we will discuss in the next section.

Major Lessons Learned

- Capacity is one of three basic requirements for production (the others being information and material).
- Capacity exists at two levels — strategic (capabilities) and tactical (units of output per period of time).
- Capacity can be measured in terms of either resources or output, with output being the most often used measure.
- Capacity is finite with respect to time; it cannot be stored.
- At any point in time, capacity can be used for production, setup, maintenance, or idle time. To increase one type of use, we must decrease one or more of the other areas.
- Capacity can be measured in terms of maximum, effective, and demonstrated. Comparisons between these help us identify the nature and source of potential problems.
- Our focus in this section has been at the micro or work center level.

Measuring Capacity

Previously we showed how there were two views of capacity — one focusing on volume and one focusing on capabilities. The two are tightly linked. Our discussion, though, did not address the issue of how to measure capacity. Measuring capacity should be, at first glance, a relatively straightforward task. What we are doing is measuring in quantitative terms the size of the volume that we can produce. It is not that simple, though. We must measure capacity not only in terms of supply (e.g., the number of hours of capacity available) but also in terms of output (e.g., the number of units produced).

To begin this process, we must identify the limiting or bottleneck resource. This could be labor or a piece of equipment or tooling. For the purposes of this discussion, we will focus our attention on labor. Having identified the limiting capacity, we must identify and calculate two different measures of this capacity. The first is the attendance hours. This is the total number of hours that the resource is available. For example, if we are dealing with a one-shift operation, then the attendance hours might be 9-1/2 hours. That is, the employees are on site from 7:00 a.m. until 4:30 p.m. Even though the employees are on site for 9-1/2 hours, they are not working that entire time. The amount of time they are actually available for use (and during which they can be expected to be used) is referred to by some as the net available operating time (NAOT).

The difference between the attendance hours and the NAOT is typically accounted for by breaks, lunch, and planned maintenance time. For example, returning to our previous example, we know that we have 9-1/2 hours of time available; however, we also know that the employees are allowed two 10-minute breaks. They are also given a 60-minute lunch break and 20 minutes for end-of-shift maintenance. This means that the NAOT is equal to 9-1/2 hours or 570 minutes – 20 minutes (breaks) – 60 minutes (lunch) – 20 minutes (maintenance), or 470 minutes. Before going any further, though, it is important that we stop to examine another time element that is often considered when calculating NAOT — safety capacity.

Safety Capacity

Safety capacity is a buffer of capacity that management often introduces when planning capacity usage. It is a percentage of capacity or a time period during which, on average, we plan to leave capacity idle. For example, in our previous example, we could have planned to assign 45 minutes of every day to safety capacity. At first glance, this seems like an absurd policy. Why should we plan to leave perfectly good capacity idle? The answer is that safety capacity exists so that it can be used to accommodate unplanned orders or unplanned situations.

The management recognizes that we are working in an environment that is highly dynamic. These uncertainties are encountered either on the demand side (e.g., the arrival of new unplanned orders from customers which require immediate processing or unplanned or uncontrolled changes to the status of existing orders, either in terms of due dates or order quantities) or on the supply side (e.g., problems with processing times or the need for extra units). One way of accommodating these uncertainties is to set aside capacity "just in case" it is needed. It is this "just in case" nature of safety capacity that lies at the heart of the controversy surrounding this concept.

While some managers believe in the use of safety capacity, others see no real value in it. Some managers do not consider safety capacity when measuring capacity and planning its use. The reason is simple. For many managers, safety capacity implicitly recognizes that management is unable to control either the flow of orders into the firm or the amount of capacity that is actually needed on the shop floor. As a result, management must turn to safety capacity as a means of protecting itself and the production system from these problems. Rather than addressing the real root problems, we have elected to protect ourselves. Instead of better educating our customers about the real need for discipline when placing or changing orders, we have decided to accommodate the customers' bad habits. We have said, in effect, that it is acceptable for customers to put in orders at the last minute and expect them to be delivered on time. In addition, other managers recognize that on average safety capacity results in dead capacity. We are paying for all the costs associated with this capacity, and these costs are real; yet, there is no real revenue planned from this capacity. Safety capacity only generates revenue when the level of capacity required exceeds the amount that we have planned for. When this might occur is random and can be difficult to predict. Whether or not safety capacity is considered when planning production must be left up to the discretion of management. However, if it is, then the implications (and costs) of its use should be considered and factored into the resulting analysis (see Figure 12).

Making the Transition from Volume to Output

Now that we have considered the issue of volume, we must now convert this volume into planned output. In doing so, we must recognize that this is a more complex process. The reason is that the amount that is produced (i.e., output) is influenced by numerous factors, including (but not limited to):

1. Product mix (the more we make of the same products, the higher the output due to the need for fewer setups and opportunities for improvements through learning)
2. Length of runs
3. Accuracy of standards
4. Past experience with the products (necessary capacity might be more difficult to determine if the product that we are building is relatively new or a prototype)
5. Stability of priorities (when priorities are changing, we may abuse capacity as we react rather than plan for the best use of capacity)
6. Scheduling system in place
7. Level of work load (how much work is waiting to be processed)

It is important to note that the output is also influenced by demands coming from the customers (ideally we want to produce products at the rate at which

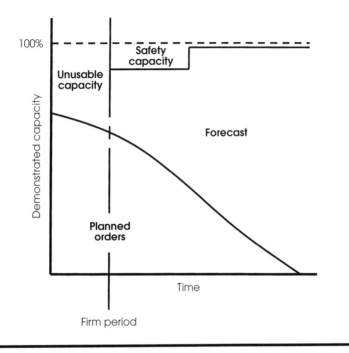

FIGURE 12. Manufacturing Load

the customers want them) and the bottlenecks within our system. Given these and other variables, we will next explore the question of how to make this transition.

Major Lessons Learned

- In measuring capacity, we must understand and work with the differences between the volume and output dimensions of capacity.
- Attendance time identifies the total time that the capacity is present but not the time that it is available.
- To identify the time over which the capacity is available for use requires determining the net available operating time (NAOT). This could be calculated as the attendance time minus breaks minus lunch time.
- A major area of controversy is whether or not safety capacity should be considered or included in the NAOT calculation. Safety capacity is a buffer that we add to protect ourselves from uncertainties in demand and processing.
- When dealing with safety capacity, it is important that we recognize that on average safety capacity is dead capacity. That is, it is capacity

that we pay for but which generates little, if any, revenue. It also deals with symptoms rather than addressing the underlying problems.

■ When making the transition from volume to output, we must recognize that this transition is influenced by numerous factors such as product mix, operator experience, condition of equipment, work loads, and scheduling practices.

Making the Transition from Capacity in Isolation to Capacity within the Processes

Overview

Before proceeding any further, let's take a moment to clear up some of the confusion surrounding the notion of safety capacity. Safety capacity exists because we, as managers, cannot always manage our customers. They possess an important resource that explains their lack of response to our attempts to manage them — money. With this resource, they have the ability to demand that things be done their way, not ours. As a result, we must recognize that there are many instances in which we have no other recourse but to use safety capacity. However, if we are to use safety capacity, we must: (1) understand its true costs and the problems it creates, (2) justify it on the basis of relevant costs and benefits, and (3) undertake a program of educating our customers so that they can also become aware of the true costs generated by the need for safety capacity. Safety capacity is like safety stock. It represents an investment. By investing in safety capacity, we implicitly forego the opportunity to employ the resources invested in safety capacity for other uses. As a result, we must allocate appropriate charges to safety capacity. These charges should reflect opportunity costs.

This said, we will now turn our attention to making the transition from capacity in isolation to capacity of the entire system. This transition will enable us to develop a greater understanding of capacity overall and the need to measure, identify, monitor, and manage bottlenecks. It will also prepare us for a discussion of the need to diagram processes. It is our contention that we cannot manage what we cannot measure or describe. We will focus on bottlenecks and constraints by discussing several interesting paradoxes, one of which is the paradox of time and capacity. That is, the steps that define the critical path (minimum lead time) may not be the same as those that define the minimum output.

Capacity in Isolation to Capacity of the System

When we look at a work center in isolation and when we can successfully measure and monitor that work center's capacity, we may think that we have gained mastery over the concept of capacity. This is not the case, however. All that we have done is gain mastery over one piece of the capacity puzzle. We have learned how to ride a tricycle; we are not yet able to drive a Formula 1 racing car. Now we will need to take a step back to look at that work center within the context of the larger process.

In most firms, the work center is part of either a larger department or process. What is a process? We define a process as "a collection of activities that transform inputs into an output that offers value to the customer." A process is a collection of activities, such as the work center previously discussed. It is important to think in terms of processes for several reasons. First, processes are central to all firms. Firms do not consist of work centers; rather, they consist of processes, which, by the way, happen to make use of the capacities provided by various work centers. These processes are everywhere. They are used in strategic planning, order entry, resource planning (the manufacturing planning and control system described in most operations management textbooks is an example of a process), product design, process design, supply chain management, and logistics management.

Second, processes define what firms can and cannot do. In other words, a process defines capabilities and limitations. To understand this concept, take a look at a fast-food restaurant — what they can and cannot do is greatly influenced by the processes that are in place.

Third, processes are described in terms of four major attributes: structure (how things are arranged), capability, time (the amount of time needed for the process to complete one unit of output), and capacity. To understand the relationship of capacity and processes, consider the process illustrated in Figure 13. Here we see four different activities linked. For one unit of output to be produced, the order must travel from activity A, through B and C, and then to D. Within this process, we are faced by two challenges. The first is to measure capacity at each activity, and the second is to measure overall capacity.

Measuring Capacity at the Activity

The first problem is really a natural extension of the issues and themes that we have been discussing up to this point. An activity can come in one of many

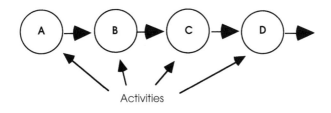

Activities

FIGURE 13. Positioning of Activities Sequentially

forms. It can be a single piece of equipment or a single worker (a shipping clerk, for example) or a tool. In this case, measuring capacity is no problem. However, the activity could also consist of several pieces of equipment working in parallel. That is, suppose that at activity B (Figure 13) there are three machines (1, 2, 3). An order entering this activity can go to any one of the available machines. The capacity in this case is simply the sum of the three activities. That is, if machine 1 has a capacity of 100 pieces per hour, machine 2 a capacity of 75 pieces per hour, and machine 3 a capacity of 125, then the overall capacity for activity B is 300 (100 + 75 + 125). Notice that in this case the machines have to be parallel and independent (i.e., what happens at 1 has no impact on operations at 2 or 3 and visa versa).

A different capacity is obtained if the three machines at activity B are organized in sequence. That is, 1 takes place before 2 which takes place before 3. Now the capacity for the activity is based on the capacity of the machine with the lowest capacity (in this case, machine 2). The issue of whether the capacity is organized in parallel or in sequence is critical because it affects all of our capacity calculations. Now we have the conceptual tools we need to evaluate the capacity of the entire process.

Calculating Capacity of the Overall Process

Returning to the process in Figure 13, we find that we have the following capacities (assuming design or maximum capacity): A, 250 units per hour; B, 300 units per hour; C, 400 units per hour; and D, 225 units per hour. This particular process is sequentially organized, so the resulting capacity of the process is 225 units per hour (based on the activity with the lowest capacity). By the way, the activity with the lowest overall capacity is the bottleneck. The bottleneck need not be a manufacturing activity. The bottleneck could be a service activity (e.g., processing an order) or it could be a design, engineering, or support activity.

Why is it important that we identify the bottleneck? The answer is that the bottleneck dictates the output of the overall process. We can make no more than the bottleneck. If we increase the capacity of the bottleneck, we increase the output of the overall process. If we increase the capacity of a non-bottleneck activity, we have spent our money without affecting overall capacity in any way. With this in mind, we are now ready to explore the challenges of managing and documenting the overall process.

Major Lessons Learned

- A process is a collection of activities that transform inputs into an output that offers value to the customer.
- Processes affect the firm on at least four dimensions: time, structure, capacity, and capability.

- To change any one of these four traits, we must change the underlying processes.
- Firms are defined in terms of their processes, not their activities.
- How the capacities of a process are measured is dependent on whether the capacities are organized in sequence or parallel.
- If the capacities are parallel, the overall capacity is the sum of the individual capacities.
- If the capacities are sequential, the overall capacity is determined by the operation with the lowest output.
- The activity with the lowest output is the bottleneck.
- Investing in bottlenecks increases output; investing in non-bottlenecks increases expenses.

Production Scheduling Is No Substitute for Capacity Planning

Overview

One of the problems that we have encountered on more than one occasion involves the question of what production scheduling system or procedure should be used. A typical encounter goes as follows. We are sitting in our office and the telephone rings. When we answer it, the person at the other end first verifies that we are indeed the people who are responsible for the "Back to Basics" series in the APICS journal, then the caller goes on to explain the reason for the phone call.

Typically, such callers want us to recommend a new or different production scheduling system. In some cases, these people have been to seminars or presentations where they have heard about a new (typically described as "revolutionary") dispatching system. The callers want us to comment on these systems or provide them with recommendations. We nearly always begin by asking them a series of questions, the first of which usually involves the issue of capacity. We ask for a description of the company's capacity planning system. Most callers, when faced with this question, often respond by telling us that: (1) they have some form of capacity planning in place (even though they cannot seem to remember the specific type of system), and (2) capacity is not the problem, scheduling is.

The problem here is often not in production scheduling; instead, it lies in capacity. Production scheduling, if it is to be effective, must have sufficient capacity for completing the schedule that has been assigned to it. If it does not have sufficient capacity, then production scheduling transforms itself. It is no longer a scheduling system; it is a rationing system. Production scheduling now becomes a device for rationing. It becomes a system for determining which orders (and which customers) will be satisfied and which orders are to be left waiting on the shop floor, late and ever falling behind schedule. Without adequate capacity, we cannot meet all of the demands placed on the shop floor. Production scheduling, in short, depends on good capacity planning and is not a substitute for capacity planning. This point is critical. It is also a point that is often lost on managers who call us with questions about better production scheduling systems. However, before we can explore the requirements for effective scheduling, we must define exactly what we mean by the concept of production scheduling.

Production Scheduling: Defining the Concept

Production scheduling has been studied under a number of different concepts — sequencing, job shop scheduling, and even shop floor control (in its broadest context). Underlying all of these terms is the notion of allocating resources over time to the performance of a collection of tasks. This idea can be restated in another, potentially more meaningful manner. Production scheduling is a *constrained matching* process. This definition focuses attention on the two critical attributes of production scheduling: (1) the process of matching, and (2) the notion of constraints.

Matching takes place along the dimensions of *resources* (what resources are needed by the orders), *time* (over what specific time intervals are the resources allocated to the orders), *quantities* (how many of the various resources are allocated to the orders), and *priorities* (what is the relative importance of competing orders). It must be recognized, however, that the matching process occurs in a constrained environment. Production scheduling is constrained by decisions made previously within the overall corporate and operations management settings relative to (1) the *capabilities* of the production system, and (2) the nature of the *planning systems* within which detailed scheduling is accomplished.

Capability is more than the production setting (e.g., job shop, manufacturing cell, assembly line, line flow, or hybrid). Rather, capability is shaped by factors such as technology, facilities (location, size, layout), work force (skill level, wage policies, employment security), and organization. Decisions made concerning these factors affect what the production system can and cannot do. An assembly line, for example, can excel at building a standard item, but it has difficulty dealing with highly customized products or production schedules subject to constant revision. Ideally, there should exist a fit between the requirements of the orders released for production and the capabilities of the production system to which these jobs are released.

Production scheduling is not only constrained by the inherent capabilities of the production system, but also by the production and capacity planning systems within which detailed scheduling occurs. The aggregate production schedule identifies the orders to be produced, as well as the times (due dates) by which these orders are to be manufactured. It establishes the demand that the detailed production scheduling process must satisfy. Furthermore, the capacity planning system determines the amount of resources that are available for meeting the demands placed on the production system. Production scheduling must operate within the limits or constraints imposed by these planning systems. It is a fact of manufacturing life that production scheduling cannot create capacity nor can it cancel orders when demand exceeds capacity. At best, production scheduling can conserve capacity and fine-tune its use. For example, when a dispatching rule groups orders to take advantage of common setups, the result is not an increase in capacity but the conservation of existing capacity.

While production scheduling is best characterized as a constrained, matching process, it has other important attributes or roles as well. First, production scheduling serves as a communication device for workers on the shop floor. An examination of a Kanban system makes this role clear. A Kanban serves as a visual signal, and its arrival indicates that someone in the process needs a replacement order. Second, production scheduling does not exist in isolation from other shop floor activities. Rather, it is part of the larger shop floor control system. This means that production scheduling is influenced by how the other activities in the shop floor control system are executed.* It is in this relationship that we see the link between production scheduling and capacity.

The Problem of Production Scheduling: Solving the Wrong Problems and Hiding Others

With production scheduling, we take the work load released to the shop floor by the planning system and create a dispatch list. This list tells us what orders to work on, where, and in what order. Often, we use production scheduling to cope with a wide range of problems. Some of these problems really are best tackled using production scheduling, but production scheduling is not appropriate for handling many of them. For example, production scheduling can be applied to situations involving changing order priorities, expediting, quality/vendor problems, or orders having less than normal lead times. In each of these cases, even though scheduling can be used, it is the wrong tool. The answers to these problems can be found elsewhere, not in scheduling; yet, we often encounter situations where managers are turning to production scheduling to deal with these problems.

This is not the only difficulty encountered when using production scheduling. In addition, we must recognize that the use of production scheduling can make production/shop floor problems invisible to management and others in

---

* For more information on this aspect of production scheduling, see Melnyk, S.A. and Denzler, D.R., *Operations Management — A Value-Driven Approach*, McGraw-Hill, Burr Ridge, IL, 1996, chap. 18; Melnyk, S.A. and Carter, P.L., *Production Activity Control*, Dow Jones-Irwin, Burr Ridge, IL, 1987.

the firm. Consider, for a minute, what happens when a scheduler or shop floor manager finds that a component that must be built next to keep the entire order on track cannot be scheduled because the vendor-supplied components have been found to be defective. One option facing that manager is to stop production until the underlying problem is resolved. The problem with this approach is that in many firms it is the manufacturing equivalent of suicide.

For many managers, a more appropriate solution (a production scheduling solution) might be to identify another order that currently has all of its components available and is ready to run. This job is then pulled forward and scheduled on the machine. Notice, though, what this scheduling revision has achieved. By moving this order forward, the work center continues to run. To management (or anyone else who is not familiar with activities taking place on the shop floor), everything seems to be normal. The equipment is running and work seems to be progressing; however, this is not really the case. By working on the jobs out of order, we have delayed one critical order (thus causing the rest of the orders to be delayed) and we have moved up another order (thus increasing work-in-process). We have also effectively hidden the problems being experienced. Finally, we are faced by the challenge of trying to explain the deteriorating performance of the shop floor.

Prerequisites for Effective Production Scheduling

Before production scheduling can achieve its objectives (the timely and cost-effective completion of jobs in a manner consistent with how the firm competes in the marketplace), certain objectives must be present. The following are the most important:

- Realistic lead times must be assigned to each order released to the shop floor for processing.
- Orders must have realistic, credible due dates assigned to them.
- There must be sufficient capacity available based on the particular period and work center.
- There must be sufficient stability in the schedules (a lack of excessive changes to the schedules) so that operations can be appropriately planned (without having to immediately replan when the schedules change).
- The logic of the production scheduling system must be transparent and logical to the people using that system.
- Particular care must be exercised to ensure that the schedules released to the shop floor do not violate the capacity of any bottlenecks.

Production scheduling operates best when these requirements are met. In the following discussion, we will focus our attention on the last of these requirements — taking the bottleneck into consideration.

Major Lessons Learned

- Not all problems dealt with by using production scheduling are true production scheduling problems.
- Production scheduling is not a substitute for effective capacity planning.
- Production scheduling is a constrained matching process that brings together load (demand) and capacity on a period-by-period basis.
- Production scheduling, if incorrectly used, becomes a rationing device. It becomes a system for doling out capacity that is insufficient to meet all the demands. It can also hide shop floor problems, thus making them effectively invisible.
- Production scheduling, if it is to be effective, requires that certain prerequisites be satisfied first. These requirements must be addressed at the planning or strategic levels of the firm.

Production scheduling is something that everyone can do but few do very well.

Don't Schedule the Shop; Schedule the Bottleneck

Overview

Previously, we have shown that dispatching and production scheduling, no matter how well done, are no substitute for effective capacity planning. Before we can allocate capacity to the various jobs, we must first make sure that there is enough capacity to ensure that no one job becomes late due to a lack of capacity. Once we have provided sufficient capacity, we can turn to production scheduling. The task of production scheduling is to generate a schedule (dispatch list) that takes advantage of the knowledge of those who work on the shop floor to maximize efficiency. It is here that we generate schedules that recover capacity by taking advantage of such factors as efficient workers and jobs that share setups. This is one reason that we can never completely computerize the production scheduling process. We need the insights and knowledge of those who work with the jobs on a daily basis. This is another reason why it is useful to remember: *The planning system is responsible for feasibility; the production scheduling system is responsible for efficiency.*

At the heart of feasibility is the issue of capacity. Before exploring this concept in greater detail (and getting to the heart of this section), it is important to emphasize that just because we have a Material Requirements Planning (MRP) system in place, we do not necessarily have capacity planning. One of the critical but often overlooked premises on which most MRP systems are built is that there is infinite capacity. This premise simplifies planning within MRP but is rarely met in practice.

At first glance, the task of dealing with capacity seems to be a daunting one. After all, every production system (be it manufacturing or service based) consists of numerous resources and work centers, each with its own capacity limits. This situation would seem to imply that we have to plan capacity for each work center or resource. This requirement would result in a very complex scheduling system; however, there is a simpler answer to this problem. Rather than planning and scheduling every work center or resource, what we should do is to focus our attention on the constraints or bottlenecks.

Bottlenecks: The Key to Simplified, Effective Scheduling

As previously noted, a bottleneck is nothing more than that resource in a system that has the highest load on it. This resource could be a resource such as labor, or it could be a machine or the office staff. In some companies, the constraint could be the design staff or the shipping dock. In other companies, the bottleneck could be externally imposed and take the form of the shipping company or one or more suppliers. Bottlenecks affect our firm's ability to produce output within the required time interval (the lead time); they also affect the level of output. In short, the bottleneck limits our ability to generate revenue and ultimately profit.

Before going any further, we should look at a situation when we are faced with more than one bottleneck. In such a situation, we have two options. We can live with the two bottlenecks, or we can choose to invest in one of the bottlenecks and thus eliminate it. Neither of these options can be addressed at the shop floor or production scheduling level. This is a strategic issue. As a result, it must be addressed at the top management level. As a general rule, we have come to the conclusion that, when faced by more than one bottleneck, it is best to invest and eliminate all of the bottlenecks except one. It is far easier to control one bottleneck than it is to control two. In fact, the effort to control bottlenecks increases exponentially with the number of bottlenecks.

While it is true that we cannot schedule ourselves or our production system out of a capacity problem, there is an option that we have available. This option begins with a simple but critical tactic — we schedule the bottleneck first (see Figure 14). We then establish the priorities for the open orders. Next, we determine the sequence for the orders going into the bottleneck. We work on ensuring that the bottleneck is always busy. This could mean that we may have to surround this constraint with inventory. The cost of having inventory is far less than the cost of having the bottleneck sit idle for the lack of work. Remember that one hour of capacity lost at the bottleneck is one hour lost for the entire system. This loss translates directly into lost revenue and profit.

Before we can schedule the bottleneck, we must first identify it, which means we must first do a load analysis, an activity carried out at the planning system level. Part of the process of identifying a bottleneck is to assess the ease

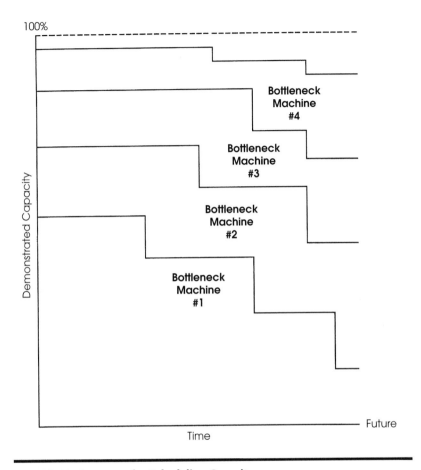

FIGURE 14. Sequence for Scheduling Capacity

with which we can increase capacity at the bottleneck. This issue introduces the debate between infinite as compared to finite capacity planning.

At the heart of this debate is the extent to which it is practical to treat capacity as infinite or finite in nature. When working under infinite capacity planning assumptions, the planner does not worry about imbalances on a period-to-period basis, as long as the load demand placed on the overall system is less than the level of resources available. This assumption makes sense when existing capacity levels can be easily and quickly increased. These increases can be obtained in many ways, including the use of overtime, part-time labor, cross-trained workers, second shifts, or alternate routings. As load increases at one work center or area, resources are moved to that area (typically away from other areas that are not experiencing such high levels of demand).

In contrast, with finite capacity planning, we assume that resources cannot be easily or cost effectively increased in the short term. As a result, when the load exceeds capacity in a given period, there is an increase in overall lead times (as orders spend more time waiting for the required resources to become available). When working under the assumption of finite capacity, the primary option available to managers in dealing with a condition of capacity overload is to reschedule the excessive work to periods when the needed resources are available. This involves either releasing work early or delaying the release of orders until future periods.

We now see that the resource that is highly loaded and has capacity that cannot be easily increased in the short term becomes the bottleneck. It is this resource that we must schedule first.

After the resource load has been determined, we can then sequence the jobs through the operation based on the lead time required to perform the operation so that subsequent tasks can be performed in time to meet the due date. Now we can sequence the jobs through this one highest load bottleneck based on priority determined by due date. Once the highest committed resource has been loaded, we then select the second highest and load that one. And so it continues through our resource capacity loading program until all job requirements have been loaded. And we mean all of them. Not loading all the demand is like never balancing our checkbook. We will never know how much of our resources are already committed; when sales asks for the next available minute of time for the next job, we won't know what to say.

This is true production scheduling. We are scheduling our resources based on the capacity constraints. We have yet another feature that we must look at, though. After the top constraint has been loaded, all other constraints will be loaded at less than 100% capacity and will have idle time available. A word of caution here. The only thing that we can use this idle time for is completing jobs that do not require time on higher loaded equipment.

In production scheduling, we are looking at the interrelationships of the jobs, the interrelationships of the parts within the jobs, the bottleneck resources, and the sequencing of jobs through the equipment to maximize total throughput.

Now we are at a point where we can really say that we are effectively and efficiently scheduling production. Keep in mind, though, that if a date has passed by without a task being done, we only have two choices. We can obtain additional capacity to finish the task, such as through overtime or subcontracting. Failing that, we have no other choice but to reschedule the late job (after informing planning and, ultimately, the customer).

Major Lessons Learned

We have looked at the details of scheduling the production system and have shown that this activity has its beginning in good capacity planning. The next step is to identify and focus on the bottleneck. Our goal at this point is to ensure that this bottleneck, once identified, is always busy. Identifying the bottleneck helps to establish a process for effective scheduling:

1. Determine equipment load by resource.
2. Determine lead time required to meet the ship date.
3. Schedule the highest loaded resource first (remembering that we should have only one true bottleneck).
4. Schedule the second highest resource next.
5. Continue scheduling until all the resources are complete.
6. Sequence the jobs within the resource based on lead time and due date.
7. If an assigned task is not completed by the due date, either get more capacity or reschedule the open orders.

7 Understanding the Nature of Setups

Why Be Concerned with Setup?

Before the advent of Just-in-Time (JIT) manufacturing, with its emphasis on short, predictable lead times, setup was viewed as a given. Setup times were often given to manufacturing people after being determined by the engineering group. This time allowance determined what resources were needed and when. When process improvements were sought, setups were often overlooked (after all, they were a given); however, this view of setups has now radically changed.

We are now beginning to see setups for what they are — a *necessary but not value-adding activity.* Our customers do not pay us to do setups. Setups are rarely, if ever, seen as being something of value to the customers; however, we need setups so that we can carry out the operations necessary to convert inputs into outputs of value to our customers. Furthermore, we now recognize that setups have major implications for the attainment of manufacturing excellence. They consume capacity and time. In addition, they can be a major source of manufacturing variability. Variability, in turn, leads to lead times that are not predictable. As a result, a task for many manufacturing systems has been to reduce setups. Yet, before we can reduce setups, we must first understand what is meant by the concept of setup. This is a challenge that is easier said than done. In reality, there is a great deal of confusion surrounding this concept.

Understanding the Nature of Setups: The Basics

What Is Setup?

Before we can define setup, it is important to note that setup has two distinct meanings. The first meaning pertains to capacity, as setup is a source of demands placed on capacity (the other usages of capacity are summarized in Figure 15). The second deals with the cost implications of setup. This latter perspective is most frequently encountered when dealing with lot-size calculations — the famous, or infamous, economic order quantity (EOQ), but in our discussion here we will focus on the capacity aspects of setup.

As previously noted, capacity is finite. That is, there is only so much capacity available in any time period. This finite amount of capacity is consumed by such demands as production, maintenance, idle time, and setup time. There is an inverse relationship between setup time and capacity availability. The more time that we take for setup, the less time is available to meet the other demands that are placed on capacity. On the other hand, the less time that is used for setup, the greater the capacity available to meet the other demands.

Given the importance of setup time, do we know what setup time is? We define setup time as the amount of time (which often translates directly to the amount of capacity consumed) that it takes to change a machine or work center to go from the last part of a production lot to the first consistently good part of the next production lot. Within this definition, the key words are "amount

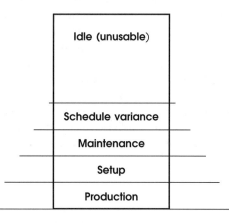

FIGURE 15. Consumers of Capacity: The Only Positive Consumer of Capacity Is Production; the Others Are Non-Value-Added

of time" and "first consistently good part." Setup does not end until we are able to consistently produce parts meeting our specifications. Any parts that are rejected because they do not meet our specifications during this initial period are considered part of setup.

Setup time recognizes that the amount of capacity consumed is subject to some degree of variation. This variation occurs because setup times are themselves the result of a process which typically consists of four major types of activities. Whenever we begin to set up a work center or machine, we must first get the necessary items, so this first step of the process consists of finding and reviewing the specifications and processing information for the part to be produced next. We must also locate the necessary tooling and verify that the tooling is able to process the necessary number of parts (e.g., there is sufficient life available in the tooling). The next step is to tear down the previous setup and rebuild the equipment (by inserting new tooling) so that it can produce the required item. The third step involves positioning the parts and ensuring that the items to be produced are correctly lined up. Finally, there is adjustment. This adjustment activity affects both the parts and the tooling. It is here that we clamp down the tooling, run a few parts to test the positioning, and reposition the tooling if necessary. We continue this process until we are sure that the tooling is correctly positioned and that we are now able to consistently produce quality parts.

We can see from this brief description that these four activities can be sources of variability. For example, if the tooling and processing information are not readily available near the work center, we would expect that the operator will be forced to leave the work center to search for the necessary tooling and information. In practice, it is difficult to predict how much time that this searching activity might take. To understand why, consider the search process that many operators seem to follow when searching for tooling and processing information (the following process description is based on our experiences and observations on the shop floor, and it can be verified by most shop foremen).

The first step in looking for tooling and processing information is for the operator to take a "pit stop." This is inevitably followed by a smoke break, for those so inclined, or a coffee break, for the others. Then, the operator inevitably feels the need to talk with people encountered while looking for the necessary items. Often these conversations cover such topics as the problems of finding the lost tools as well as the status of last night's football game (everywhere except in Cleveland, that is). Eventually, the process sheets and the tooling are located. The operator can now proceed back to the work center (again at his own pace). However, this description of the search process is only appropriate if the tooling and/or the process sheets are ready to be used (e.g., the tooling is in good condition and the process sheets are up to date). If there are problems with any of the items to be located, then it is almost impossible to predict the total time needed to find the proper items with any degree of certainty.

Internal Activity, External Activity, and Capacity Usage

At this point, you might be wondering whether there is any relationship between setup time and capacity usage. The time needed for the four activities previously described does not necessarily consume capacity. To make the link, we have to introduce the concepts of internal and external activities. An internal activity describes any activity that must be done while the machine is sitting idle (not making parts). Internal activity reduces capacity availability. In contrast, an external activity describes any activity that can be done while the equipment is working, producing parts. Setups that take the form of external activities do not reduce capacity.

We can now see that the impact of setups on capacity usage is strongly dependent on the percentage of time accounted for by internal activities. For example, suppose that the total setup time is 150 minutes (2-1/2 hours). Of this total time, internal activities make up 120 minutes, while external activities consume only 30 minutes. Under these conditions, we can see that one setup reduces capacity by 2 hours (120 minutes) because the equipment must be stopped and idle while the setup is taking place.

Why Worry About Setups?

There are several reasons for being concerned about setups. First, the internal activities of setups can affect capacity availability. Second, setups can be a source of variability (i.e., unpredictability and uncertainty). In addition, when faced with large setups, most managers react by increasing the lot sizes or run quantities. Finally, it is important to remember that setups are fundamentally non-value-adding but necessary activities. That is, a setup is necessary because it enables us to produce those parts that the customers want. It is the act of producing the parts that is the real source of value to the customer. Customers normally do not pay for setups. As a result, we are obligated to keep setups to a minimum and to control any variance present in them.

General Guidelines for Managing Setups

To manage setups, there are certain basic things that we must do. We must first study the existing setups without judgment. This means that we must study what is being done without criticizing or judging. Often this requires documenting the setups using video cameras. Next, we must break the setup process down into the four steps previously described. In addition, we must identify for each step the distribution of time, which means identifying not only the mean time but also the variance. Having done this, we can categorize the setup times as being either internal or external activities, after which we can then focus on: (1) identifying the sources of variability and reducing or eliminating them, and

(2) converting internal activities into external ones and then reducing the amount of external setups.

Understanding the Nature of Setups: Setups and Lot Sizing

We have been focusing on one specific component of capacity utilization — setup times, and we saw that there was a strong relationship between capacity utilization and setup times and that this relationship strongly depended on the degree to which the setup times consisted of internal activities. Internal activities include any activity that must be done while the machine (or work center) is sitting idle. We also found that setup times could be a major source of variability. Our focus will now turn to setup as it pertains to lot sizing. For most practitioners, this is a major source of confusion and misconceptions.

Lot Sizing: A Review

Before we look at the role of setups in lot sizing, we must first understand what is meant by lot sizing. Lot sizing is the amount or quantity that we order or make whenever we need to fill an order or replenish inventory (and the existing amount on hand and on order is insufficient to cover the demand). This amount can range from making just enough to meet the immediate demand to making enough to cover a number of periods of demand. This amount can be determined in various ways. We can base this amount on factors such as the size of the storage bin (or the number of storage bins) or experience-based rules of thumb (whenever we place an order, we order 125 units because that was the largest demand that we had encountered in the past). Alternatively, we can use a more analytical method to determine this amount. Most of these more analytical procedures are built around the relationship between the holding costs and the setup costs. Understanding this relationship is necessary to understand the role of setups.

If we do not have any inventory on hand when a demand occurs, then we incur a cost in the form of setups. For every order that occurs under these conditions, we must have a setup. Under many conditions, this can be a very expensive and time-consuming situation. For example, if we have very large setup costs (e.g., $1000 every time we prepare the equipment or place an order with our supplier), then such an approach does not make a great deal of sense when we are building only a couple of units costing $5 each. One way of avoiding this situation is to hold enough inventory to meet demand (and avoid the setup costs); however, the larger the amount of inventory, the greater the costs of holding inventory. Under these conditions, we are faced with the problem of determining the optimum order quantity when considering both total holding costs and total setup costs.

This problem is not a new one. It was first addressed by Harris in 1913. At this time, Harris was working at General Electric and developed a formula which is now referred to as the economic order quantity (EOQ). What he determined was that the most economical order quantity was the one that balanced the total costs of holding costs with the total costs of setups.

Setup Costs

As we pointed out previously, setup is the amount of time that it takes to change a machine or work center or to place an order (in the case of purchasing) to go from the last part of a production lot to the first consistently good part of the next production lot. This definition has to be modified slightly when it is applied to lot sizing. In this case, a setup cost is comprised of the *variable costs* associated with going from the last part of a production lot to the first consistently good part of the next production lot. The concept of *variable costs* is critical. Setup costs are limited to only the actual variable costs that are incurred. Every time a setup takes place, these costs are generated.

Setup Costs: An Example

A young production supervisor was interested in determining the setup costs so that he could calculate the order quantity for a specific part that ran through a certain machine. He called a colleague in accounting and asked if he could look into the matter and call him back with an answer. About a week later, he got a call from his friend in accounting who told him that the setup costs were $510. This number shocked the young supervisor. The supervisor asked the accountant to explain this number. Simple, was the response. The $510 was based on the following costs: (1) $150 came from two setup men who had to work two hours each on the job (at $37.50 per hour, including all benefits); (2) $10 came from the machine operator who had to help with part of the setup; (3) $150 came from tooling expenses, of which $50 was directly consumed by the setup and $100 was a preparation charge (this number was determined by taking the total overhead expenses for the tool room and dividing by the number of orders processed); (4) $100 came from the budgeted scrap of four units that were expected to be discarded as a result of the setup (each unit had a variable cost of $25); and (5) $100 was budgeted allocations.

There is a basic problem with this setup cost, though — it is overstated. It is overstated by the presence of fixed costs and allocated costs. These costs should not be considered when determining the setup cost for lot sizing. For example, the first $150 should not be considered because the setup men represent fixed costs. We would pay them for eight hours regardless of whether they do setups or not. The tooling expense represents a mixture of allocated and variable costs. The $100 is an allocated cost and should be discarded. Consider this cost for a moment. If we had experienced one more order in the tool room and if this $100 was really a variable cost, then the costs should go up by $100. But this is not actually what would have happened. Remember that we arrived at this number by taking the total overhead for the tool room and dividing it by the total number of orders. The arrival of the additional order, then, would have reduced the preparation charge, as total expenses/(orders + 1) < total expenses/(orders). Similarly, the last $100 should also be dropped because it is a budgeted allocation. The "true" setup cost should have been $10 (for the operator) + $50 for consumed tooling + $100 for scrapped parts, or $160.

What this little exercise shows is that determining the actual setup costs is really a more difficult exercise than simply accepting numbers provided by accounting. At this point, we should introduce the concept of relevant costs, because they are central to this discussion. Simply stated, relevant costs are any costs that would change in response to an action or decision or any costs that would affect the outcome of the decision. Irrelevant costs, in contrast, are unaffected by the decisions or actions. This difference is critical because it forces us to recognize that, in many firms, many costs assigned to setup costs are irrelevant.

Before we leave this discussion of setups and relevant or irrelevant costs, there is one more point that should be raised — setups and relevant costs are both capacity sensitive. For example, a setup cost might be irrelevant at a low level of capacity utilization (e.g., 50%) but highly relevant at high levels of capacity utilization (e.g., 85% or greater). The reason for this difference in treatment is because of the opportunity costs of the capacity. Under conditions of low capacity utilization, the capacity consumed by the setup would have been largely left unused. As a result, using these resources would not raise the variable costs incurred. In contrast, when the capacity utilization increases, there is now a real cost incurred by using capacity for setup. Consuming one hour of machine or labor time for a setup would leave one less hour for production. To do the setup, we would have to sacrifice production, which ultimately is where we generate value for the firm. In this case, setup would represent a real loss, creating a relevant cost.

The notion that relevant costs are capacity sensitive explains an interesting response taken by firms when faced with a situation where setups are large and capacity utilization is high. They tend to do fewer rather than more setups. At first glance, this tactic would not seem to make sense. After all, under conditions of high capacity utilization, there are probably more orders (thus requiring different batches and setups) than can be realistically built. Under this scenario, it would make sense to build a bit of everything just so a portion of all the demand is satisfied. However, this tactic is ultimately self-defeating. When we build small order quantities of each item, we see more and more capacity consumed by setups, and capacity is the critical resource. By building large quantities of only a few items, we devote less and less capacity to setup and more to actual production. This does mean that some orders will be unfilled; however, the reality is that we cannot fulfill everyone's needs — we simply do not have enough capacity. One method of avoiding a situation in which we are building larger and larger order quantities under conditions of high capacity utilization is to work on setup reduction.

Major Lessons Learned

- Setup costs consist of only those variable costs associated with an order. Not included in setup costs are allocated costs.
- Setup costs should contain relevant costs.
- Relevant costs are affected not only by decisions or actions but also by the level of capacity utilization.
- When capacity utilization is high and available capacity very low, an appropriate tactic for responding involves that of making fewer but larger orders. This tactic attempts to reduce the amount of capacity consumed by setups.
- One way of avoiding the need to make fewer but larger orders is to focus on setup reduction. This approach requires reducing the amount of capacity consumed by setups.

8 Inventory: The Most Misunderstood Corporate Asset

Inventory Accuracy: What's So Important About the Status of Our Inventory, Anyway?

Overview

We all have inventory in our operations and have lived with inventory ever since man first began to make more than one of anything. Inventory record accuracy was not an issue to us even as recently as a few years ago because we always had inventory, and we always talked about the accuracy of our records but did little to improve the record accuracy. Inventory provides a cushion or insurance policy to protect us from poor inventory record accuracy. We are beginning to think that one way to protect ourselves from problems arising from poor inventory record accuracy is to just put more inventory in our warehouses to compensate.

Why Inventory Accuracy

So what is the big deal about how accurate our records are, anyway? We have the situation covered with our inventory. Why bother with record accuracy? In the past, when we saw that we were running out of something, we just bought more. And if we found that we still ran out of stuff, we bought even more. No big deal, right? Not so. What our accounting friends have told us is that inventory is a fixed investment that generates no revenue for us. The accountants have always been after us to lower the amounts that we have in inventory so they could spend that money elsewhere (something about investment opportunities or profit, or something like that). But isn't inventory something that we need to stay in business? Without any inventory, we would be in a position of not being able to supply our customer. Where would McDonald's be if they did not have any inventory of hamburgers to give us when we came in the restaurant? Isn't the same thing true at your company? Don't we really need inventory to be in business?

Yes, we do need inventory of some sort to stay in business but not as much as we had in the past because now we have systems that can plan our operations and eliminate the need to carry as much inventory. By lowering the inventory that we have at our disposal, though, we have less protection from the unknown. Let's take a look at how things have changed. Inventory is a buffer against the unpredictable. It protects us from variations in our ability to manufacture, as well as variations in our customers' ability to determine demand and give us an accurate forecast. Inventory also protects us from supplier quality and delivery problems, which are very important to us in light of today's demands for quick delivery.

What about our inventory record accuracy, though? There are four things we need to do correctly in today's manufacturing arena. We need accurate bills of materials, accurate route sheets, excellent capacity definition, and inventory record accuracy. We need accurate bills of materials to tell us what we need to meet the order, and we need good route sheets to tell us how to assemble that product and how much time we will need to do the work. It is also important to define our demonstrated capacity and the capacity we have available to complete the order.

The Need for Record Accuracy

An accurate inventory allows us to answer the first question we are likely to ask ourselves when receiving an order: " How much do I already have?" Inventory record accuracy is the fourth input to our operational system, and it is the second one that we go to when the order comes in to answer this question.

It is difficult to attain a high level of inventory record accuracy if our input records are poor. Input records, the bills of materials that specify the kind and quantity of materials required to build a particular product, are key to inventory record accuracy. If we do not know how much of each item we need or what the items are, then we have a problem. If the quantity is in error, we might have more on hand than we thought when it comes time to build the product, which will not delay a shipment, of course. On the other hand, though, we might actually have a smaller quantity of the units than is shown on the bill of materials, which could result in a stock-out. Both of these situations reflect inventory record accuracy problems and are undesirable.

Be aware that you can really shoot yourself in the foot when managing the inventory. Let's assume that we really do have an accurate bill of materials and have acquired all the materials that we need for a job. Then we discover that we misidentified a particular part. It is here somewhere, but we can't find it because we labeled it with the wrong name when we stored it. We end up with a stock-out and cannot ship the product. You might think that this is a problem that will not affect you, because you bring your materials in just in time and directly to the assembly bench, and they cannot be lost in the stockroom. This might be true most of the time, but without an accurate system it is possible that you may not be aware that certain materials came in. They could end up lost somewhere on the assembly bench because your system was not told that the materials had arrived — same problem, different location. You have it but don't know it, and you find yourself once more on backorder.

Another way we can make a mess is to put the materials in a different location than what the records indicate. Now we have two problems instead of one. The materials we want are not where we think they are; they are somewhere else. We will most likely find that the materials that were supposed to be in that location are now in the spot where we were looking for the lost material. Wrong locations generally create two record errors, and, with tight order windows and minimum inventories, we again have a backorder.

What we are trying to say is that mishandled inventory or the receipt of just-in-time materials can be very costly. The real cost here is not the additional inventory that must be carried to cover the errors in our system, but it is the cost of the late or lost order because we could not ship. I have seen people manage inventory to such a finite level that they were managing plastic bags used to ship the final product in two-week increments. They only had a two-week supply of bags on hand. Total annual usage was $150 worth of bags. Their records were poor because they did not think they needed good record accuracy for something as cheap as a plastic bag, and it was not on the bill of materials to order the bag in the first place. How costly was it? What do you think that the cost of this many airfreight shipments of plastic bags was per year? And, beyond that, what about the cost of delayed shipments due to any stock-outs that occurred through the year? By the way, the average order for these people was $15,000, and when they ran out of bags an average of ten shipments would go on backorder.

Inventory record accuracy is not for you? Think again. If you want to be quick and responsive to your customer, you must have a good inventory record system. And it must be accurate.

Major Lessons Learned

There are four data input points to the inventory record accuracy system:

1. Bills of materials
2. Route sheets
3. Demonstrated capacity
4. Inventory records

One of the best methods that we can employ to increase our record accuracy is to provide precise, detailed information on the input side of our materials system in these four areas.

By the way, here is what we could do with the plastic bag problem. The bags are not worth much anyway, so it makes sense to buy a year's supply of items like that, put them on the shelf, and forget about them. That way there is only one purchase order per year, and the worst we can do is have just one stock-out per year. But, we do not want more than one year's worth of stock on hand. A forecast on anything over one year is too risky in regard to obsolescence and use rates. The real cost of more than one stock-out per year would be much greater than all the plastic bags in the world. We should not spend so much time managing such small dollar values. Think about it. Accuracy is the key to maintaining a good inventory system.

If Inventory Is a Waste, Then Why Study It?

If we were to talk to any manager (be it a manager of manufacturing, marketing, or accounting) about the major problem facing manufacturing, the one area that most would cite would be inventory. For most, the current inventory levels are too high, not the right type, in the wrong place, or simply out of control. Numerous articles and books have been written in an effort to improve control over this resource; yet, in spite of this volume of material, many mistakes are still being made. In part, this situation reflects a fundamental confusion over inventory, what it is, why it exists, and what can be done to better manage it. This lack of understanding can be best demonstrated by the problem of managing inventory through FIAT (management decree).

The Fallacy of Inventory Management through FIAT

You are a production and inventory control manager in a manufacturing firm. One day, in response to a demand from finance (who sees your inventory levels as being too high), you receive a management mandate to reduce overall inventory levels by 30%. Furthermore, you must achieve these reductions within the next 6 months. You know that this objective must be met — either by you or your successor.

You review the items before you. You know that you have slow movers and fast movers. In an ideal world, you would focus your attention on the slow movers. However, you know that, given the time constraints you are facing, by the time the reductions in the slow movers took place, your successor would benefit from your actions (why do you think that they are called slow movers?). As a result, you decide to focus on the fast movers. Furthermore, to meet the overall goals of a 30% reduction, you know that you have to reduce the fast movers by more than 30% (to compensate for the slow movers). You identify and implement the changes necessary.

Because you have not addressed the underlying reasons for the excess inventory, however, you find that the reductions you have introduced are starting to create distress throughout the firm. Certain orders are experiencing stock-outs, customer complaints are increasing, and marketing is expressing their dissatisfaction. You are approached by several sales and marketing people who want to know what is happening. You patiently explain that you have had to reduce inventory levels in response to a management mandate. Eventually, after enough pain has been endured by marketing and the firm as a whole, the costs of reducing the inventory levels are deemed to be too high, and the old inventory levels are restored. In some cases, a little bit more is ordered "just in case."

With this approach, nothing in the way of a long-term improvement in performance has been achieved. Yet, we can raise an interesting question: "How do we really reduce inventory?" That is the challenge addressed in this section. The problem with inventory management by FIAT is that it attacks inventory, rather than addressing the reasons for inventory. As a result, such an approach is ultimately doomed to failure. The only way to avoid making similar errors is to understand inventory and why it exists.

Understanding the Nature of Inventory: Functions

Overview

In some ways, inventory has received a bum rap. Ever since the advent of the Just-in-Time revolution, we have been taught to view inventory as an "evil." Inventory is something that must be avoided at all costs. Yet, this is too simplistic a treatment of this corporate asset. What most managers do not seem to understand is that inventory is often a symptom, seldom the problem. Stated in another way, inventory is a residual. In general, because it is a symptom, we find that we can seldom attack inventory directly. Inventory exists because it is a response to the need for a certain function that it provides. It is this need that we must attack, not inventory. This position seems to confuse many managers. Because of this fact, we have decided to provide a thorough exploration of inventory. We think that you will find the resulting discussion both interesting and insightful.

What Is Inventory?

What is inventory? Let us start with addressing this very basic but important concept. Accountants will tell us that inventory is an asset. It is something that the firm owns. On the balance sheet, inventory is treated as part of the liquid assets of the firm. What this means is that inventory can be converted into money relatively easily and quickly, but this is not always the case. We know, for example, that obsolete inventory is not as liquid. Often, its real value is very different from its book or accounting value. The Just-in-Time (JIT) experts will tell us that inventory is an evil. Inventory consumes resources and hides problems. As a result, inventory should be avoided. This is a blanket statement and not always true. There are times when inventory has real value. For example, if we are preparing to go on vacation and the car's distributor shorts out, then we are very pleased when the dealer has a replacement distributor in stock.

Inventory is simply the positive difference between supply and demand. It occurs when over a period of time we make more than we consume. Inventory may also be an asset for which we have title (ownership) but not possession. For example, the software package coming to us from a mail-order house in California is inventory. We own it but we do not have it in our hands.

This positive balance occurs for either unintentional (we forecasted more than was actually sold) or intentional reasons. Our focus here is on the intentional reasons. Inventory occurs when the costs associated with *not* having inventory exceed the costs of having inventory. Still, why do we really need inventory? Because of the functions that it provides.

Functions of Inventory

We do not have inventory simply because we like inventory *per se*. We have inventory because it helps us cope with a problem or remediate the effects of a situation. We buy the function provided by inventory. To understand the functions provided by inventory, we turn to a scheme that was developed by one of the greatest operations management teachers of our time — Bob Britney of the University of Western Ontario (London, Ontario), who developed the following taxonomy as a way of teaching his students the reasons for inventory.

Transit

The transit function occurs because of geographical differences between the places where demand and supply occur. It occurs, for example, when our supplier is located in San Jose, CA, and we are located in Falls Church, VA. When the inventory leaves the supplier's plant in San Jose, we have title to it. The amount of transit inventory is a function of three factors: (1) distance, (2) quantity, and (3) method of transportation. If we want to eliminate this function of inventory, we can invest in a faster mode of transportation (e.g., replacing rail shipments by air), or we can look for a supplier who is located closer to us.

Decoupling

Decoupling occurs primarily within manufacturing facilities. It occurs because there is an imbalance between the rate at which products are supplied and consumed. For example, we could have two machines, A and B. A feeds B. A produces 200 units per hour; B consumes 150 units per hour. These differences could be due to problems in preventive maintenance. They could also reflect inherent differences in machine rates. Without inventory between A and B, we would have to run machine A for 45 minutes to every one hour of machine B's production rate; however, by keeping inventory between A and B, we decouple (separate) A from B. A now produces to inventory; B consumes from inventory. To eliminate this inventory, we would have to balance the production and consumption rates. This may require establishing different criteria for purchasing equipment (we would now stress system balance as compared to getting the most output for the money from each machine). Alternatively, we could invest in improved preventive maintenance. Unless the underlying reasons for the imbalance are identified and remedied, we are hard pressed to eliminate the need for inventory.

Lot Sizing

This function typically occurs because of large setup costs. When we have large setup costs, we do not find it economical to produce in lots of one unit. Rather, we produce in quantities where the average setup costs equal the average holding costs. To avoid lot sizing, we must reduce setup costs. This is the reason that JIT experts focus on setup reduction. It is necessary to eliminate the lot-sizing function. With the lot-sizing function eliminated, we can think in terms of lot sizes of one (the goal of JIT).

Anticipatory

This function occurs when we expect changes in the conditions of supply. These conditions refer to either the availability of products in the near future or the price at which they are offered. For example, we would tend to invest in this function of inventory when we expect a strike at our supplier or when we think that our supplier is about to increase the price or reduce the allotment of product that we are to receive. To reduce this function of inventory, we have to either lock our suppliers into long-term contracts or identify more stable sources of supply.

Seasonal

This function exists because of imbalances between supply and demand. These imbalances should be of a regular and predictable nature and exist when demand occurs year-round while supply is limited to one brief period of production (as in the case of blueberries in Michigan and Wisconsin). Alternatively, the demand may be limited, but the supply occurs year-round (as in the case of a toy company). Inventory becomes the method for coping with this imbalance. To eliminate this inventory, we must either coordinate supply and demand (e.g., sell the product only when the supply is available, as in the case of blueberries) or try to lengthen the demand period, as a toy company would try to do for demand for toys.

Buffer

The final function of inventory refers to inventory that is held because of uncertainties. These uncertainties could be in demand (either in quantity or timing). They could also be internal, as in the case of scrap rates. Finally, they could also result from supplier problems. For example, the lead time for our suppliers might be highly variable. Alternatively, we might have a situation where the actual quantity delivered is variable. This occurs in cases where the amount delivered is based on weight not piece units. In each case, we must first identify the source of uncertainty and eliminate it before we can reduce buffer inventories.

Applying the Functions Approach

With this approach, we have a better way of attacking inventory. The first step is to classify inventory according to the function that it provides. Once we have identified the various functions, we can examine these functions to identify the underlying root causes. It is these root causes that we must attack and eliminate before we can hope to eliminate the associated inventory. This is why inventory is a residual, a symptom that we can never attack directly.

Major Lessons Learned

- Inventory is a residual or a symptom.
- Inventory occurs whenever the amount produced over a period of time exceeds the amount consumed.
- Inventory may be an asset over which we have title but not physical ownership.
- Inventory exists because it provides one of six functions: transit, decoupling, lot sizing, anticipatory, seasonal, and buffer. Each is driven by its own set of root causes.
- The underlying root causes must be attacked to eliminate the need for the associated inventory.

Understanding the Nature of Inventory: Forms and Positions

Overview

To study inventory, we begin with a simple but important observation — inventory is a residual. We acquire inventory because of the function that it provides. To eliminate the need for inventory, we must first identify that function and then attack the need for the function. Once the need for the function has been eliminated, then we can successfully eliminate the inventory; otherwise, we find ourselves in the traditional vicious cycle of inventory described above. We can never mandate an effective long-term reduction in inventory unless we identify and attack the reasons for inventory; however, identifying the functions provided by inventory is not enough. Next, we must examine the form and position of the inventory. The form describes where in the process the inventory is located. The position of the inventory describes the location of the inventory in the supply chain.

Forms of Inventory

In general, inventory can take one of five forms, the first of which is raw materials, which refers to inputs acquired from our suppliers. Raw materials are brought into the firm. These items could be raw inputs such as iron ore or processed components such as microcomputer chips. This inventory is the most flexible in that it can be transformed into any end item that we want; however, to make this transformation requires the relatively longest lead time.

The second form of inventory is work-in-process, which includes items being worked within the process. Compared to raw materials, they are closer to the ultimate form of the product, but they are still not ready to be shipped out.

The third form is finished goods or shippable goods, which are the end items that we provide our customer. Once made, these items require the least amount of lead time (relative to raw materials); yet, they present an interesting trade-off in that they are the least flexible form of inventory because they have been configured into their final form (it is difficult to change a car painted red to one that is blue). Of the first three forms of inventory, they tend to be the most expensive (because of the addition of materials, labor, and other associated costs incurred during the transformation process).

These first three forms of inventory are the ones with which most of us are familiar; however, there are two additional forms that inventory can take. The first is maintenance, repair, and operating (MRO) inventory, which refers to the stock of items that we need for the operation and maintenance of the transformation process. In MRO, we tend to find items such as rags, oil, staples, nipples for glue guns, and gloves. In many firms, this is a major investment.

The final type of inventory is money, which indicates that management has recognized that they need not invest in inventories of parts. They can decide instead that it is in their best interest to keep the resources in cash.

With these five forms, we can examine the form distribution of our inventory (i.e., what percentage is in raw materials, work-in-process, finished goods, MRO, and cash); however, identifying these forms is not sufficient. We need to look at where in the supply chain the inventory can be located.

Position of the Inventory in the Supply Chain

In general, we can position our inventory within the supply chain at one of four locations. The first location is at our suppliers or within our purchasing system. The second location is within our plants at the manufacturing system. The third position is in the logistics system, which includes inventory that is in transit (i.e., on trucks or in rail cars), as well as inventory that is located at our various warehouses. Finally, the inventory could be located in the marketing system near our customers.

Putting Things Together

We now have three sets of dimensions. These dimensions consist of function (six different types), form (five different types), and position (four different positions). This gives us 120 possible combinations. By being aware of these various combinations, we can now begin to examine our inventory and to assign the inventory to one of the 120 different combinations. We can now begin to see the distribution of inventory. With this information, and by using Pareto analysis, we can target those combinations that represent the largest investments of inventory. By the way, if we find ourselves with inventory that cannot be assigned to any one of these 120 different combinations, then we have inventory that is out of control. That is, it exists, but we do not know why we have it or what function or position or form it is taking.

Major Lessons Learned

- It is dangerous to manage inventory through management mandates.
- Inventory can take one of five forms: raw materials, work-in-process, finished goods, MRO, and cash.
- Inventory can be located at one of four places in the supply chain: with our suppliers or in purchasing, within the manufacturing system, within the logistics systems, or in marketing near the customer.

- With these six functions, five forms, and four locations, we can develop a tool (a matrix) for determining where a particular inventory is based on its functions, current form, and location.
- The matrix can be used to direct attention and to identify those combinations that are high and that offer management opportunities for systematic inventory reductions.

Understanding the Nature of Inventory: ABC Analysis

Part 1

Overview

Having identified the forms, functions, and position of inventory, we are now in a position to begin developing some form of control over this inventory. The method that we will discuss here is the ABC system of inventory control. We will first introduce the concept and logic of ABC analysis, then we will focus on variations of this approach and how it can be used to improve our control over inventory.

Develop Categories of Inventories

One of the problems with modern computer software is that it enables us to treat all inventory equally; that is, we can look at every item in inventory in as much or as little detail as we want. This equality of treatment is not always the most effective or efficient method of managing inventory. It confuses the trivial many with the vital few. It gives the trivial many the same degree of worth as the vital few. As a result, we tend to manage all the items in inventory equally poorly. This leads to an important question of how to distinguish the vital few from among the trivial many. To do that, we must turn to a procedure that was introduced before the widespread usage of computers. This technique, ABC analysis, was created as a way of helping managers simplify their task of inventory management by focusing attention on the vital few.

ABC analysis is a system for categorization. The concept of a categorization system is not new to us. That is what we have been focusing on in this chapter when we discussed the various forms, functions, and locations of inventory which provide categories into which we could assign various portions of inventory. ABC analysis is an application of the Pareto principle, which was named after an Italian economist who, when studying the distribution of income in Italy in the 1800s, determined that 20% of the population owned about 80% of the wealth. This 80/20 rule has been found to be applicable to a number of situations, one of which is inventory control.

The Nature of ABC Analysis

The goal of ABC analysis is to assign all inventory to one of three categories (we have seen some firms use anywhere from three to eight or more categories of ABC, but for the most part three categories is a good starting point). Each category brings with it different requirements in terms of the degree of control and requirements. In general, we use two attributes when describing the categories. The first is the percentage of part numbers or stockkeeping units (SKUs). The second is the percentage of inventory value accounted for by that category of inventory. Inventory value is defined in terms of two traits — the quantity demanded per period of time (usually a year) and the cost per unit. Typically, we express the three categories as shown below:

Category	Percentage of SKUs (%)	Percentage of Inventory Value (%)
A	15 – 20	75 – 80
B	30 – 35	10 – 15
C	50	5

The A items are few in number (no more than 20% of the total part numbers); yet, they account for a very large percentage of the inventory value. The As and Bs together account for 50% of the part numbers and 95% of the inventory value. The Cs account for the remaining 50% of the part numbers but no more than 5% of the inventory value. The As are the vital few, while the Cs are the trivial many. This might now lead to the question of why we should categorize the inventory in this way. There are several reasons. The first is for inventory control purposes (to determine how closely we manage the movement of inventory). The second and often overlooked reason is for inventory reduction purposes.

ABC and Inventory Control

Once we have assigned the various inventory items to one of these categories, we have now simplified the manner in which we control the items. The A items are controlled tightly. Every time that they move from one area or person to another, responsibility for that item must be formally transferred. As a result, with A items it makes sense to have a system of perpetual inventory control. The A items are also prime candidates for a locked stockroom. If we have a cycle-counting program in place, then we want the cycle for A items to be the shortest. With A items, we want someone to be always responsible. In contrast, for C items loose control is sufficient. For example, we can decide to place C items out on the floor where they are readily available for use. This "grocery" approach to inventory management makes sense for C items. We may even draw a line on the C item bins which would indicate the reorder point. That is, if the quantity of the C item is above the line, we do not place a replenishment order, but once the quantity falls below the line, we place an order. The key rule with C items is that we never stock-out of them. The B items fall somewhere between the As and Cs.

ABC and Inventory Reduction

Many managers are familiar with ABC analysis for inventory control. Few, however, use this system for inventory reduction. One of the options available to any manager is to try to reduce the number of SKUs. As the number of SKUs increase, we need more and more powerful inventory management systems. With fewer SKUs, we can get away with simpler systems. This leaves us with the question of how to go about reducing SKUs. One way is to rely on ABC analysis. We can focus on the C items, and within this category we can look at those items for which there is little or no demand. These items consume the same floor storage space as other items for which there is a greater demand, but these items generate little or no offsetting revenue. For these items, the options facing us include eliminating them, redesigning other items to take over the functions provided by these items, subcontracting the responsibility for them to our suppliers, or keeping them. If we decide to keep them, then we must accept the costs associated with storing them.

Doing ABC Analysis

Doing ABC analysis is relatively simple and straightforward. It consists of the following steps:

1. Identify the unit cost for each SKU. This cost can be obtained from accounting.
2. Obtain the usage over a period of time for each SKU. The period of time used is usually a year. The usage figures can be the usage for last year or they can be forecasts, but in either case the analysis is done on a unit basis.
3. Multiply the unit cost for each SKU by the periodic usage to obtain the periodic dollar usage (units times dollars).
4. Sum the periodic dollar usages for all SKUs.
5. Divide the periodic usage of each SKU by the sum obtained in step 4, and multiple by 100 to get a percentage usage.
6. Rank order the percentages obtained from step 5 in descending order.
7. Identify the A, B, and C items using the guidelines previously identified (15 to 20% of the SKUs accounting for 75 to 80% of the value are As, and so on).
8. Modify the categories for qualitative considerations (we will discuss this in the next section).

Completing the following table will help you understand this process:

SKU	Unit Cost ($)	Annual Usage (# units)	Annual Cost ($)	Percentage (%)	Rank	ABC Class
1	0.05	50,000	___	___	___	___
2	0.11	2000	___	___	___	___
3	0.16	400	___	___	___	___
4	0.08	700	___	___	___	___
5	0.07	4800	___	___	___	___
6	0.15	1300	___	___	___	___
7	0.15	17,000	___	___	___	___
8	0.20	300	___	___	___	___
9	0.09	5000	___	___	___	___
10	0.12	400	___	___	___	___

Major Lessons Learned

- Not all items are equally important.
- We should focus on the vital few and manage those really well, rather than being distracted by the trivial many.
- One way of identifying the vital few is through the use of Pareto analysis, which is based upon the general principle that 80% of the problems/value can be attributed to 20% or fewer of the issues/items.
- We can implement Pareto analysis through ABC analysis.
- ABC analysis is initially based on the value of each inventory item (SKU) as defined by the combination of its unit price and periodic usage. The higher this figure, the more important the value and the closer we should monitor it.
- ABC analysis classifies every item or SKU in inventory into one of three categories — A, B, or C (with the As being most important and Cs being least important).
- The A items are monitored very tightly. The general rule for C items is that we never stock-out; we manage these items fairly loosely.
- ABC analysis can be used to identify potential candidates for elimination.
- There is a simple procedure for carrying out ABC analysis that can be easily implemented by computer using a spreadsheet such as Excel®.

Part 2

A Review

Underlying the basic and simple technique of ABC analysis are some very important issues. The first is that all items are not equally important; some items are very important and should be monitored carefully, while other items are not very important and should never be stocked out. The basic method for identifying these items is through the use of Pareto analysis. This technique relies upon the general principle that 80% of problems can be attributed to 20% or fewer of the items or issues. One basis for identifying importance is to focus on each item's value as defined in terms of the percentage of the total dollar usage for which this specific item accounts. To determine this number, we multiply the annual quantity usage times the cost per item (and then sum across all items to get the total dollar usage).

Doing ABC Analysis: The Answer

The first item of business is that of providing an answer to the problem assignment given on the previous page:

SKU	Unit Cost ($)	Annual Usage (# units)	Annual Cost ($)	Percentage (%)	Rank	ABC Class
1	0.05	50,000	2500	38.6	2	A
2	0.11	2000	220	3.4	5	B
3	0.16	400	64	1.0	7	C
4	0.08	700	56	0.9	9	C
5	0.07	4800	336	5.2	4	B
6	0.15	1300	195	3.0	6	B
7	0.15	17,000	2550	39.4	1	A
8	0.20	300	60	0.9	8	C
9	0.09	5000	450	6.9	3	B
10	0.12	400	48	0.7	10	C
Total			6479			

In this case, we see that there are 2 As, 4 Bs, and 4 Cs. Is this the end of it? Not really. This analysis is based strictly on numerical or quantitative analysis. Ignored are such factors as:

1. Lead time (the longer the lead time, the more closely we must monitor the item, thus a C item might become either a B or A item)
2. Volatility of design/engineering changes (the more frequent the changes, the more closely the item must be monitored because of the threat of obsolete inventory)
3. Storage/special requirements (the more extensive the storage requirements, the more likely it is that we will have to monitor the item more closely)
4. Perishability (the shorter the shelf life of the item, the more closely it will have to be monitored)

Any of these factors (to name a few) could cause an item to go from being a C to being either a B or an A item.

Introducing Costs

To this point, we have only focused on the demand side of inventory. What has been ignored is consideration of costs. When dealing with ABC analysis, the major cost is that of the storage space occupied by the items, which, in reality, is an opportunity cost. The space that each item occupies could be used to store other, "more valuable" items or it could be used for production. One method of assigning this cost is to examine the space occupied by each item and the cost per square foot (this number can usually be obtained from accounting). Multiplying the two together yields the cost of physically storing that item in inventory. We next take this number and subtract it from the item's annual dollar usage. The result is a number which can be either positive or negative. If positive, the number tells us that the annual usage exceeds the costs of storage. If negative, we know that the annual storage costs are greater than the annual usage. So what? Well, if the resulting number is negative, then we are faced with an item that costs us more to store than its usage justifies. In such cases, we might want to explore either dropping the item or having our suppliers (if possible) take responsibility for managing it. In any case, we have identified a candidate for elimination or rethinking.

Major Lessons Learned

■ We have learned how to carry out an ABC analysis.
■ We have learned that the category determined reflects both quantitative and qualitative considerations.
■ We have shown that by including costs of storage into the analysis, through the assignment of costs for floor space, we can identify candidates for elimination, rethinking (i.e., redesign), or assignment to a vendor for control.

Managing the Slow Movers Using ABC Analysis

Overview

As we have learned, we can identify and focus our attention on the critical few A items; however, there is another aspect to consider — what to do with the slow movers. Here we will identify several different strategies that can be used to deal effectively with this problem.

The Problem of the "Slow Movers"

The slow movers are those items or SKUs for which there is very little or no demand or for which the demand is highly sporadic. Typically, these items fall into the C category of the ABC analysis. As we have previously noted, these are the items that constitute about 50% of the part numbers but only account for about 5% of the total value (where total value is based on total revenue and demand usage). Yet, we must realize that a single C item, while generating less demand per unit, consumes about the same amount of overhead resources. These resources include those used to store the item (physical space, stockkeeping personnel, obsolescence costs, and the time and effort to enter and maintain the data in the computer system), as well as those resources required to maintain demand for the products that use these items. For example, we need to have marketing personnel who can forecast demand, and we must consider the time and effort needed for an engineer to design and revise products to maintain this demand. At some point, we may consider that, for many of our C items, these costs often exceed any benefits that we may gain.

Before going any further, this statement raises an interesting but important concern — if this is the case, then why aren't more managers aware of this imbalance between costs and benefits? There are several possible reasons. The first is that few managers or systems even collect all of the costs incurred in supporting a given component or SKU. These costs, if not recorded, are assigned to overhead. Once assigned to overhead, they lose their managerial usefulness. They are now part of a large cost component for which no one is responsible. If these costs are not recorded, then they cannot be managed. It is important that we never lose sight of the critical management dictum — *we cannot manage what we do not measure.* The second factor is the fear of not having inventory or components in stock. Every marketing manager or salesperson seems to fear the effects of dropping a product from the catalog or of reducing inventory levels, because they might lose one or more critical customers. The third and final factor involves the measurement system. Put bluntly, most production systems do not reward activities that reduce the number of SKUs (they do indirectly through cost, but not directly). As a result, the measurement system focuses our attention elsewhere in the process (and not on the total number of parts in stock).

Managing the Bottom 5%

One way of gaining control over the C items is to turn the logic of ABC analysis around. ABC focuses our attention on the critical or vital few — the 15% of the SKUs that generate or account for about 80% of the sales. Now, though, instead of focusing our attention on the top 15%, one effective tactic is to focus our attention also on the bottom 5% (while simultaneously *not* ignoring the top 15%). These bottom 5% items are ones for which there is very little demand (but potential costs). If these bottom 5% can be eliminated, then we can expect to see a potential improvement in the overall performance of the system.

These bottom 5% should be reexamined on a regular basis (e.g., once every quarter) to eliminate the potential "dogs" — those products with no demand but high costs. Yet, before we can eliminate them, we must recognize that there is a process that must be followed. The first step in this process, of course, is to flag or identify these bottom 5% items using a procedure such as ABC analysis. The second is to examine them using a cross-functional team (including representatives from marketing, finance/accounting, production and inventory control, product design, and manufacturing). This team is charged with evaluating every slow mover and justifying the continued presence of a slow-moving component.

Strategies for Managing Slow Movers

At this point, it is important to recognize that we have several well-defined strategies for dealing with slow movers. The first is that of dropping the SKU entirely. With this approach, we declare the part obsolete. This approach makes sense when there are no other parent assemblies that use this specific part. It also makes sense when there are no outstanding demands (service, replacement part, or otherwise) for that specific part. Finally, this approach makes sense if top management is willing to accept the costs of writing off obsolete inventory (a major reason why more firms do not do this is because of concerns over who will bear the resulting costs).

A second approach is that of transferring the responsibility from production and inventory control which can occur in one of two ways. First, we can transfer the responsibility to a supplier, at which point it becomes their obligation to provide sufficient availability. In this sense, the item has now become a vendor-managed inventory (VMI) item. When we transfer it out, we must recognize that we have converted the costs from overhead to variable, and we have eliminated the demand for internal space generated by that item.

A third approach is to transfer responsibility from production and inventory control to marketing. If marketing (or some other group) insists on stocking a really slow mover, one option (which requires top management support) is that of making marketing responsible for stocking and managing the component.

The fourth option is that of redesigning another component so that it can be easily adapted to provide the same production functionality provided by the existing component that we want to eliminate. This strategy has been long followed in the "white appliance" industry (firms that manufacture dishwashers, washing machines, ovens, etc.).

Another strategy is that of red-tagging. This procedure is most commonly associated with the "5 S" program frequently found in Kaizen events. This procedure attempts to identify and visually label (with a red tag) all unnecessary items. By doing so, our goal is to flag them so that they can be easily and quickly moved out. We are now in a better position to organize the remaining necessary items. With red-tagging, we practice the principle of "if in doubt, throw it out." We move the tagged items out to a separate area. In their place, in the main stockrooms, we only leave the red tag itself. This tag contains such information as the part number, where it can be located, and the time that it was last used. Typically, if the part is needed, the red tag is pulled and given to someone in inventory control. That person is then responsible for getting the part from the secondary location. If such a request is generated, then we have evidence that there is still demand for the item (and we have evidence that this part should not yet be disposed of). However, after some predetermined period of time (e.g., 6 months), if we still have no demand, then we can now eliminate the physical stocks. We have firm evidence that there is no real current demand for it. Dropping this stock does not mean that we purge all our systems of information about this item. We keep the data (drawings, master information, etc.) so that we can build it again in the future, should the need arise.

The final strategy is the default strategy. This is the strategy of still keeping it in stock and of accepting the benefits (and costs) of having this part in stock.

By using these various strategies, we have a better, more structured, and more comprehensive process for dealing with the slowest movers. Ideally, if we can eliminate these slowest movers, then we can potentially reduce the number of parts that we must control. With fewer parts, we have fewer items to control and, as a result, lower costs, fewer resource requirements, and simpler systems, which is a true "win/win" situation.

Major Lessons Learned

We have focused our attention on the bottom 5% as determined by ABC analysis. These items are often ignored by many managers. By focusing on these really slow movers, we recognize the potential for the total costs of these items to outweigh any associated benefits. When dealing with these items, we need to review them on a regular basis, using a cross-functional team (we need all of the insights and perspectives brought by the members of such a team). Finally, when dealing with these really slow movers, we must recognize that we have several strategies available to us:

- Eliminate the part from the system.
- Transfer the responsibility for the item from our department to a vendor or to someone else in the firm.
- Redesign another part so that it can satisfy all of the functional requirements provided by the specific slow mover (thus allowing us to drop it).
- Red-tag the product.
- Keep the item (default strategy), and accept the costs and problems associated with its continued presence.

With these strategies, we now can recognize that the task of production and inventory is more than simply managing the fast movers or the parts that generate most of the value. It also involves the question of appropriately managing the very slow movers, the bottom 5%.

From Inventory Accuracy by Accident to Inventory Accuracy by Design: Key Concepts

Overview

One of the basics of manufacturing excellence is that of inventory accuracy. Inventory accuracy is critical. We cannot expect to have an MRP system run correctly with inaccurate inventory records (otherwise, we get the famous results of GIGO — garbage in, garbage out). We cannot expect to operate an effective and efficient formal production and inventory control system with anything less than very high levels of inventory accuracy. If we do not have inventory accuracy, then over time we can expect to see increased inventory (to protect ourselves), stock-outs, frustration, increased expediting, emergence of numerous caches of hidden inventory, and the presence of an informal system (where who you know, not what the system tells you, is important). Without inventory accuracy, we cannot answer that most basic of questions that we always encounter when receiving an order, "Do I have any of this on hand?"

While most people recognize the importance of inventory accuracy, few seem to understand how to achieve this goal. Many managers seem to think that the key to inventory accuracy is one thing, such as new technology. What these managers do not seem to recognize is that inventory accuracy is the result of a system, not any one factor operating in isolation. That is, we want to develop a system that makes inventory accuracy an inevitable result. This is the notion that we will be exploring here.

The need for a system to generate inventory accuracy, rather than relying upon a single component (such as technology) was driven home some time ago when we were called in to help the physical plant department of a Big Ten university improve their level of inventory accuracy.

The manager of the physical plant stores had called to ask if we would be willing to review the current inventory control that was in place. In this university, the physical plant was responsible for maintaining all the stores needed for the maintenance of facilities on the campus, which included areas such as classrooms, offices, and dormitories. Dormitories were a major concern, as they always seemed to require repairs (especially on Monday). Attempts to do physical inventory counts had shown management that the accuracy of the inventory records was extremely poor (running at just under a 70% accuracy rate for the most accurate records). The manager responsible for these stores had to take action, and quickly. This was the situation that we found ourselves in when we went to visit the Stores department on a Monday — the busiest of the days.

When were taken on a tour of Stores, we noticed that the Stores department was run using a "grocery store" approach. That is, anyone who needed parts would enter Stores and take a picking cart and record form. They would pick the items that they wanted and then proceed to a checkout, where the items taken would be recorded and appropriate adjustments made and entered. That is the way that the system was supposed to work. On this particular Monday, though, that was not the way the system was actually working.

On this day, chaos reigned supreme in Stores. Maintenance personnel, spurred by numerous crises (such as an elevator that did not work in one of the tallest dorms), were running in and out. As they passed through Stores' stocks, they took not only what they needed but also a bit more (just in case). As they ran out of the storage room, they paused long enough to tell the clerks that they did not have enough time to stop and get the forms processed and that they would get them "on the way back." Of course, this never happened.

By the end of the tour, it became evident that the only way for Stores to maintain good inventory accuracy was through a periodic physical count. After the tour ended, everyone returned to the manager's office, where the real purpose of the tour was revealed. The manager had just returned from a trade show in Chicago, where he saw a bar-coding system in operation which promised to be the solution to all of his problems. By having a bar-coding system in place, all anyone had to do was to pass a wand over items and accuracy would be maintained. All that was needed was a justification for the fairly sizable investment. Unfortunately, the manager was quickly brought back to earth when he was informed that the bar-coding solution would not solve his problem. For one thing, it entailed more work compared with what was being done currently (which was nothing). What was needed was an overall system for developing and maintaining very high levels of inventory accuracy.

Defining Inventory Accuracy

Inventory accuracy can be defined as the extent to which the physical count (what we find in the bins and on the floor) agrees with the counts found in the computer records. The computer count is critical because it is what is used by systems such as MRP for planning requirements and for scheduling production. It is used for determining whether or not we can address the question asked earlier — "Do I have any of this on hand?" It is also important to understand the trade-off between record accuracy and inventory. If we want to avoid inventory, then we need to plan better and more precisely. That, in turn, requires record accuracy. If we have poor record accuracy, then we cannot avoid inventory, which is a major cost of lacking inventory accuracy.

Inventory accuracy can be measured in one of two ways — in monetary terms and in physical terms. On the one hand, there is a great attraction to using dollars as the unit of measure for inventory. This is one unit that everyone in the firm understands. It also allows us to compare different items of inventory (each of which has its own unit of measurement, such as inches vs. units vs. pounds); however, dollars should never be the primary unit of measure. The reason is that costs we pay for inventory change over time, and we also use inventory over time. This results in our having to choose what cost we should use in valuing our inventory. Typically, we can use either FIFO (first-in, first-out) or LIFO (last-in, first-out). With FIFO, we assume that we always use the first units that we obtained first. As a result, we should evaluate all inventory at the latest costs. With LIFO, we assume that the last units put in were consumed first. With this logic, we value all units at the earliest costs. What this means is that we could have the same units of inventory in stock, but the dollar measurement could vary depending on which logic we are using.

If we should not use either LIFO or FIFO, then what should we use? The answer is to remember that production is driven by units, not monetary counts. We needs units to complete production and to evaluate the extent to which the schedules are feasible. As a result, we should measure and evaluate inventory accuracy in terms of physical counts, but we can use inventory measured in monetary terms as a secondary measure of accuracy. The latter should never be the primary measure of accuracy — that measure must always be based on quantities.

Determining how to measure accuracy, by itself, is not enough. We must add another dimension — location. In other words, inventory accuracy is evaluated in terms of quantity *and* location. We have to know the count by location. Without location, all we have left is "four walls" inventory control (that is, the total count is available somewhere within the four walls of the plant). With this type of inventory control, we must go out and find the inventory, thus leading to the "biblical" method of inventory control (seek and ye shall find) followed by the Christopher Columbus method of inventory control (discover and land on it). The act of searching for inventory should be recognized for what it is — waste.

The next task is to determine the minimal acceptable level of inventory accuracy. The general answer is that, at a minimum, we should be able to maintain 95% inventory accuracy. Ideally, we should develop systems that enable us to approach 100% inventory accuracy. Are there any firms where inventory accuracy is not only the goal but also expected? The answer is "yes" — banks and drug stores come quickly to mind. How many of you would be willing to put your money in a bank that advertises 95% inventory accuracy? What are the implications of 95% inventory accuracy? To assess this issue, consider the following table:

Number of Components in End Item	Accuracy per Item	Overall Inventory Accuracy
1	95%	95%
2	95%	.95 × .95 = 90%
3	95%	.95 × .95 × .95 = 86%
4	95%	.95 × .95 × .95 × .95 = 81%
5	95%	.95 × .95 × .95 × .95 × .95 = 77%
6	95%	.95 × .95 × .95 × .95 × .95 × .95 = 74%

As we can see from this table, 95% accuracy results in a reduced probability of having all of the items we need in stock. The only way of avoiding this spiral is to increase the level of inventory accuracy.

Before leaving this discussion, we should deal with the issue of tolerances. Without tolerances, physical counts are required to agree *exactly* with the record counts. This is not always possible nor desirable. In cases where we do counts using a scale, then variations in the products themselves or resulting from such factors as the weather can cause the weight counts to differ for the same number of items. Second, we should recognize that the ABC classification previously discussed can be used to structure our tolerance structure. That is, we would like to see the tightest tolerances for the A items and the loosest for C items. A small variation in an A item can create a large variation in value; in contrast, large variations in a C item will create only minor variations in the total value of the inventory. In general, the following provides an example of how we might structure the tolerances:

Category	Tolerances
A	±0.5%
B	±2.5%
C	±5%

With this approach, we consider the inventory records to be accurate if the actual amount is within half a percent of the record count. If the error exceeds this tolerance, then we have an inaccurate record. Whenever there is a difference between the physical and record count, one widely used procedure is to adjust the record count to the physical count. With this approach, we now have one simple method of measuring and reporting inventory accuracy. When we report our inventory records as being 98% accurate, then what we are saying is that the physical counts for 98% of the items are within the tolerances set; the remaining 2% are not.

Major Lessons Learned

We have shown that inventory accuracy is critical if we want to be able to reduce our reliance on inventory and if we want to make sure that we can keep promises made to our customers (both internal and external). We have shown that inventory accuracy is a quantitative measure primarily stated in physical units of measure which reflects the degree of agreement between the physical quantities found on the shop floor and the quantities listed in the records (paper or computer). Inventory should always be measured in terms of unit count by location. Further, to simplify the process of assessing fit, we can and should use tolerances (the amount of disagreement between the two counts that we are willing to tolerate before we view the inventory counts as being out of sync with the records). Finally, we learned that 95% is a bare minimum level of inventory accuracy for seeing good results coming out of our production and inventory control system (MRP or other). With this base, we can now proceed to the challenge of developing a system for making high inventory accuracy inevitable.

From Inventory Accuracy by Accident to Inventory Accuracy by Design: Developing the System

Overview

Now that we understand the reason why we are so concerned about inventory accuracy and the key definitions and concepts regarding inventory accuracy, we are ready to examine the key elements of the structure and the process by which we can ensure that high inventory accuracy is an inevitable outcome.

The Structure for Inventory Accuracy

There are five elements in any system designed to ensure high inventory accuracy: (1) the locked stockroom, (2) accountability, (3) measurement, (4) simplification of counting, and (5) the cycle-count. The notion of a locked stockroom is relatively straightforward. All items of inventory should be under some form of control. In some cases, this control can take the form of a physical stockroom (a room surrounded by barriers and with a limited number of entrances and exits). In other cases, it can take the form of a line drawn around the inventory. At the heart of the locked stockroom is the fact that the items in stock are under control. This was the problem with the stores situation described earlier. Ultimately, no one was responsible for the inventory; there was no real locked stockroom. Before moving on, there is one additional point that should be noted. If we decide that we want to release inventory to the shop floor so that it can be used and picked as needed, then we should use a reorder-point technique. We can reduce the inventory quantities by the amount of the released quantity (over which we now have no control). In the bins on the shop floor, we can use a two-bin system or mark the bins with a reorder line (when the quantity of that item drops to the line, place another order). When the floor places an order for a replenishment quantity, then we would issue the quantity to the floor and deduct it from our records. We can maintain an idea of usage by means of backflushing.

Accountability, the second element, requires that someone is held accountable for inventory accuracy who realizes that part of his or her performance evaluation is based on how well inventory accuracy has been maintained. Poor inventory accuracy will affect this person's performance reviews. Further, the people assigned this responsibility must be given complete authority over their areas. They can do whatever they see appropriate, as long as their actions do not compromise the overall ability of the system to complete orders in a timely and cost effective manner. By empowering these people, we place the authority and responsibility where it should be — at the source.

The third element is measurement. We must have a system of measures that can be used to assess inventory accuracy (one method of providing such a set of metrics is described in greater detail later on). The measures should be quantitative, easy to understand, and directly reflect the ability of the people to attain high levels of inventory accuracy. If possible, these measures should be publicly posted. This action achieves one major objective — it makes performance highly visible. If the people and the stockroom are doing poorly, then everyone knows, just as they know if they are doing well (which eliminates rumors of poor inventory control). These people now have an incentive to improve their performance.

The fourth element, simplification of counting, refers to how items are stored. In many systems, we see boxes of parts belonging to the same part number. Often, these boxes contain different unit counts. In one plant we visited, we saw five boxes containing the same castings. The first box had 42 units; the second, 38; the third, 37; the fourth, 41; and the fifth, 40. The problem with this arrangement was that every time that this part number had to be counted, all of the boxes had to be opened and counted individually.

One method of avoiding this problem and of simplifying the process of counting this item is to standardize the containers. That is, each of the boxes now contains 40 units which are arranged in five layers of eight castings each. The rule that we now follow is that we will only have one box open at any point of time. When a box is opened, we will use the items layer by layer. We will only go on to the next layer in the box when all of the items on the current layer have been exhausted. With this arrangement, the process of counting inventory is greatly simplified. We begin by counting the number of full boxes and multiply by 40; for the open box, we look at the number of full layers and multiply each layer by 8, and we only really count the pieces remaining in the layer that is currently being drawn from. With this approach, we can now physically count an entire stockroom in less than half a day.

The final element, cycle-counting, is the critical element. It is at the heart of an effective system for ensuring high levels of inventory accuracy.

An Effective Cycle-Counting Program: The Key to High Inventory Accuracy

We want to give you a tool here that will not only provide you with a higher level of inventory record accuracy but will also lower the cost of maintaining that level of accuracy and keep your operation in business. Remember that when we shut down our operations to take a physical inventory, we lose production. With an effective cycle-counting system, our operation keeps working while we perform our count.

The first element in an effective cycle-counting program is the assignment of responsibility. Ultimately, it should be the people in our stockroom who are responsible for cycle-counting. The reason for this assignment is fairly straight-forward — they are the ones who are the most knowledgeable as to what our materials look like and what the part numbers are, and they are the ones whose lives will be simplified by having inventories under control. All is not free in this world, and cycle-counting is no different; it does come at a cost, but the cost savings can be immeasurable.

Before discussing the structure of this system, it is important that we understand the pros and cons of both the physical inventory method of checking our inventory and cycle-counting. First, we will look at the drawbacks of the physical method of checking inventory:

- No correction of underlying causes of inventory errors
- Numerous part identification mistakes
- Shutdown of plant and warehouse for inventory
- No long-term, permanent improvement in record accuracy

Now let's take a look at the positive aspects of a well-structured and correctly implemented cycle-counting system:

- Timely detection and correction of causes of error
- Fewer mistakes in part identification
- Minimal loss of production time
- Systematic improvement of record accuracy

Cycle-counting is basically very simple. Every morning, a portion of the inventory is counted. The cycle-counter is given a list of parts to count and all available information about those parts *except* the number of parts that the records show are in inventory. Why? Because if we send someone out to find 1675 units of a specific part in our inventory bin, guess how many parts the cycle-counter is likely to find — 1675. If it is an easy count (few parts), the cycle-counters will do it, but why tempt fate with large numbers of parts to count? Instead, we should have stockroom people do a daily count. After the count is complete, each of the physical counts can be cross-checked against the records to see if there is a count match (or if the count is within tolerances). If there is not a match, here is a sequence of actions to take:

1. Total the counts for all locations.
2. Perform location audits.
3. Recount.
4. Check to account for all documents not processed.
5. Check for item identity:
   a. Part number
   b. Description
   c. Unit of measure
6. Recount again, if necessary.
7. Investigate error factors:
   a. Recording error
   b. Quantity control error
   c. Physical control problem
8. Investigate reasons for both positive and negative errors.

Now that the count is complete and we know the reason for the errors in the records, we can just change the records. Right? Wrong. Now that we know the reason for the error, we want to correct the cause of the error so that it will not happen again. But this sure sounds like a lot of work that we do not want to do right now.

By the way, how many times a year do you count your inventory items? Generally speaking, to meet IRS requirements, we must count our entire inventory at least once a year. So, that is what we do with only the C items. In general, we want to count our B stock twice a year and our high-moving, high-dollar items (A items) at least six times each year. It sounds like we have added a lot of work, but we really haven't. Let's look at the workload for the people in a hypothetical stockroom. We will assume that this stockroom is charged with the management of some 10,500 parts that are kept in our stockroom. The following workload table shows how much activity is involved in the cycle-counting (C/C) approach. It also compares this activity with the level of activity demanded by a physical inventory count.

Inventory Class	Number of Items	C/C Counts (per year)	Total Count	
			C/C	Physical Inventory
C	8000	1×	8000	8000
B	2000	2×	4000	2000
A	500	6×	3000	500
Total inventory counts per year			15,000	10,500

As can be seen from this table, the workload did go up by requiring an additional count of 4500 parts per year, but take a look at the workload. Cycle-counting should be done every day. If our operation works 5 days a week, 52 weeks a year, then we would work 260 days a year. If we cycle-counted 15,000 parts per year divided by 260 days, then we would have to count only 58 parts per day. That's not a lot. If we have a stockroom of this size, we probably have more than one attendant (most likely three, one for each shift). Now we are looking at less than 20 parts per person per day. Not much of a workload here, and the workload gets even less.

When we do a physical inventory, we must count all the items at the same time. As such, some of the bins are full and some are empty; on average, they are half full, so we are counting an average volume of inventory items. When we cycle-count, even if we count some items more than once a year, we can choose when in the year we will do the count. A good time to count a bin is at the point when it is either empty or nearly empty (this is the so-called quick cycle-count). There are several advantages to this approach. First, it is quick to count such a bin. Second, such items are most likely to be the cause of a potential stock-out. Third, it makes the personnel very aware of the real costs of poor inventory accuracy. Finally, it encourages the people involved to iden-tify the causes of poor accuracy. As a result, over time, count accuracy goes up and the workload goes way down. Following is a list of considerations for when to cycle-count that can save further time and money and minimize inconve-nience to the production operation:

- Count when the bin record shows empty.
- Count at reorder points (also verifies the need for the order).
- Count on off shifts when no receipts are processed.
- Count in the early morning just after the MRP has been updated and parts have not been pulled for the day's operations.

- Count when a bin record shows less than needed for an upcoming job.
- Count C items at the slow point of the year.

Take a look at the first item again. Why on earth would we want to count a bin that our inventory records show is empty? Because that is probably where we misplaced that last shipment of parts that we have not been able to find.

After we have gone through all this and have found the errors of our ways, it is finally time to correct the inventory bin record. Accountants may not like this because we are always changing the value of our assets, but this can easily be handled with a variance account. At least temporarily, we will need to continue with the physical inventory. Generally speaking, we will need to verify to our auditors that the cycle-counting procedures are better than the physical inventory method. It usually takes two physical cycles to establish credibility, at which point we can stop taking the physical inventory.

There are several cost savings realized by the cycle-counting method due to:

- Elimination of the physical inventory
- Timely detection and correction of inventory errors
- Concentrating on problem-solving
- Development of inventory management specialists in the stockroom
- Fewer mistakes
- Maintenance of accurate inventories
- Reinforcement of a valid materials plan
- Less obsolescence
- Fewer excesses
- Elimination of inventory write-downs
- Correct statement of assets

A final word on the physical inventory — when is our inventory record most accurate using the physical count? Naturally, the record is at its best the day after the physical count has been completed, but its accuracy declines from that point on for the rest of the year. And the list of benefits of the physical inventory procedure is very short. Thus, the savings generated by taking the physical inventory include:

- None

Major Lessons Learned

- There are five elements necessary to build an effective system for high inventory accuracy: (1) locked stockroom, (2) accountability, (3) measurement, (4) simplification of counting, and (5) cycle-counting.
- Cycle-count is always preferred to a physical inventory count.

- All cycle-counts should be done by the people who work in the stock-room.
- Never overlook the need to count those bins that the computer system says are empty.

When the elements described here are successfully implemented, we will find ourselves with a system for which high inventory accuracy is not simply an accident but is inevitable.

9 Odds and Ends of Manufacturing Basics

Manufacturing Excellence Draws on Multiple Sources

Up to this point, we have focused on certain central themes; however, these themes, while critical to manufacturing excellence, are not sufficient by themselves. There are other elements that contribute to manufacturing excellence. In this chapter, we have brought together all of these other elements of excellence, and our attention will be directed primarily to two major areas: part numbering and bills of materials. These are two areas that are central to every production system — be it in manufacturing or the services. For many firms and managers, however, these two elements are poorly understood. As a result, they have generated a great deal of confusion in many systems, especially as firms undertake the upgrading of their Manufacturing Resources Planning (MRPII) systems or have made the move to Enterprise Resource Planning (ERP) systems. It is hoped that the following discussion will clarify what are essentially critical building blocks of manufacturing excellence.

Part Numbering: An Introduction

Overview

Recently, we received an interesting call from a small firm (less than 20 full-time employees and less than $3 million in sales). One of the production managers wanted to have us come in and visit their firm for a simple "look-see." However, after talking with the manager for some 15 minutes, it was obvious that there was a deeper rationale for the visit. The firm had three managers, each of whom was responsible for one of three major areas of the business. Recently, the managers had begun to examine the part-numbering scheme that was in place. What they saw did not excite them.

The current part-numbering system was a mess. It had evolved over time, with no real thought or logic behind it. Instead, the part numbers reflected the approach taken by the person assigned to purchasing or the development of the bill of materials. At times, this task was done by a temporary employee; at other times, it was done by a secretary; at still other times, it was done by a planner or an engineer. Some people used vendor part numbers, while others developed descriptive part numbers (a 3/4-inch copper nut became a 75nutcu) or other approaches. After some 35 years of business, the production managers decided that it was time to take stock of their system for part numbering. They found that there was no consistent logic to the part numbers. The part numbers developed and used in one area of the business bore no resemblance to part numbers used in either of the other two areas. As a result, it was possible for one area to issue an expedite order for a part only to find that there was an abundance of that part in one of the other two areas. Furthermore, the managers were convinced that they could not use the part numbers to consistently identify the underlying nature of each part. As a result, this situation was generating a great deal of heated discussion.

One of the results of these discussions was that the general manager (who was second to the owner in authority) decided that something had to be done. When he attended a trade conference, he saw a software package for vendor part-number management. To him, this was just the ticket. Without any input from anyone at the plant, he bought it and returned to the plant to present the staff with his done deed. He expected thanks, but instead the outcome was a great deal of confusion and frustration. He thought that he was giving the plant manna from heaven; some of the staff looking at this same product were convinced that it was the manufacturing equivalent of pigeon droppings. This was the situation that we walked into when we visited the plant.

After touring the plant and being introduced to four of the key staff people, we listened as the staff began to debate the relative merits of various software packages and the need for a new part-numbering system. After this had gone on for about 30 minutes, it became evident to us that the key issue in this discussion was not really software, but rather part numbering. We stopped the discussion and asked a question that puzzled the personnel present: "What is a part number, and what is its role?" The following is based on the resulting discussion and comes from the realization that there are many managers out there who share the same confusion over this very basic but critical concept.

Defining the Concept of Part Numbering

To define the concept of a part number, we must address the three separate but interrelated questions of what, why, and how. The first question deals with what a part number is, the second addresses the issue of why we should worry about part numbers, and the third deals with the various approaches for assigning a specific part number to a specific item.

The "what" of part numbering is relatively simple to address. A part number is simply a unique identifier assigned to any component or part. The key trait of a part number is that it *uniquely* identifies the part. That is, to every part there is only one part number. That part number differentiates that part from any other part. It establishes the unique nature of the part; it gives the part an identity. That is the major task of a part number. This part number can be numerical (e.g., 123456), alphabetical (e.g., ABCD), or some combination of the two. As long as the part-numbering scheme achieves its objective of uniquely identifying the part, we really are not that interested in how the part number is assigned. Note that no number can be assigned to more than one part, and no part can have any more than one number assigned to it. This feature is critical, as we will see in our discussion of the "how" of part numbers.

The second question (why) is somewhat more complex. Part numbers are important because of the numerous roles and functions that they fulfill. Part numbers are required if we are to use and implement a system of bills of materials. We need a part-numbering system because we need a way to identify a part so that we can refer to it. Second, part numbers facilitate communication and greatly reduce confusion. For example, we can ask for a screw; however, there are many items that can be classified as a screw. Some screws are long, some short; some are made of steel, some out of other metals (such as stainless steel or copper); some are designed to be used in metal, while others can be used in wood. Alternatively, we can ask for a 1234CUV, which could describe a 3/4-inch copper screw intended to be used in steel. The part number, in this case, acts as a form of shorthand. In its designation, we capture succinctly a great deal of information.

Part numbers are required if we are to carry out any activities such as cycle counting, inventory management, and ABC analysis. There is another function facilitated by a part number that has caused a great deal of confusion in the past — engineering design control. From past studies, we know that anywhere upwards of 20% of all parts in the typical firm are either duplicates or near-duplicates of other parts currently in the firm's inventory system. These duplicates or near-duplicates represent waste. If we design again a part already in the firm's system, then we are reinventing the wheel. We are investing time, resources, and effort to design a part that we already have. Such resources would be better used to work on truly new designs or to improve the manufacturing or purchasing characteristics of existing designs. In addition, we should be able to learn from past designs and use the lessons and experiences when designing or costing new designs. For example, suppose we are in the business of designing pens. If we are simply taking a basic black pen and changing the outside color (to gray, perhaps), then we should be able to base the bulk of the costing and processing information on the black pen.

In the past, the reason that most firms and their management were unable to deal with these problems and opportunities was because they lacked the tools and procedures. Inevitably, they relied on the memories of the people involved to provide these linkages and identify these opportunities. Such an approach is inherently flawed for several reasons. The first is that people forget. The second is that there is no assurance that the same person who was involved in the design of one part will be assigned to the design team involved with the second, duplicate part. The third is that there is nothing inherent in the system that will flag such duplicates (or near-duplicates) or that will help users identify near-duplicates. The solution to this dilemma, for some, has been the strategy of assigning significant part numbers.

Assigning Part Numbers: To Be Significant or Not To Be Significant

When assigning part numbers, we must recognize that there are two basic approaches. The first is to use nonsignificant part numbers. Under this approach, the part number, by itself, contains no critical product information. The numbers can be randomly assigned (which is essentially the method used by the firm described previously) or the numbers can be sequentially assigned. When a part number is sequentially assigned, each part number is based on the last part number assigned plus one. If the last number assigned was 123456 and we have a new part, that part gets the number of 123457 (123456 + 1).

Alternatively, we can use a form of significant part numbering. With this approach, some or all of the digits found in a part number contain selected information about the part, its manufacture, its source, its dimensions, or its processing. At first glance, this approach is very attractive because it provides the designers with a disciplined method of determining whether the new part is truly new or whether it is nothing more than a duplicate (or near-duplicate) of another part in the firm's part database. Similarly, the use of a significant part-numbering system allows us to identify parts of interest so that we can learn from their processing or costs. The key to these benefits lies in the part-numbering system, which becomes the vehicle for search and description. With a significant part-numbering system, we can deconstruct (break apart) a part number into its components and search for particular features. We can also build a new part number. Typically, this is done through a software system that prompts the user to respond to a series of questions pertaining to critical traits of a part (e.g., whether or not this product is obtained from a supplier). After these questions are answered, the software package delivers a part number. At first glance, this seems like a very attractive approach; however, as we will see below, there are some critical problems associated with a significant part-numbering system.

Major Lessons Learned

- Part numbers are critical and fundamental building blocks of any production and inventory control system.
- A part number, in its most basic form, does nothing more than uniquely identify a part. For every part number, there must be only one part, and for every part, there must be only one part number.
- A part number can consist of any combination of letters and numbers.
- A part number is important because it is required for most production and inventory control activities, including bill of materials structuring, ABC analysis, and cycle-counting. More importantly, a part number provides an important method of reducing confusion and eliminating ambiguity when it comes to describing a specific part or item.
- One major feature of a part-numbering system is that it should enable users to describe important traits of a part or allow users to search among existing parts to look for designs having certain traits of interest. This has led to the emergence of significant part-numbering systems (as compared to nonsignificant part-numbering systems).

Understanding Part Numbering: Significant Numbering (Yes or No)?

Overview

In the production and inventory control world, part numbering is one of those basics that many people agree we must do, but few people seem to know how to do it well. We will now turn our attention to a specific aspect of the part-numbering problem — significant part numbers.

Reasons for Using Significant Part Numbers

At first glance, the use of significant part-numbering schemes seems to be something that is truly logical and self-evident. With significant part numbering, we see a part number as being nothing more than a series of key traits. For example, assume that we need a part number for an item that is purchased domestically and is used in an engine that our firm produces. The part is metal, cast, and an A-class item. We can capture these traits within the part-numbering scheme using the following logic:

Part Number Position	Values
1 (source of part)	1 — Internally manufactured 2 — Manufactured by a wholly owned supplier 3 — Supplier-provided
2 (nature of part)	1 — Current production 2 — Prototype 3 — Test part
3 – 4 (part descriptor)	En — Engine-related Br — Brake-related Tr — Transmission-related
5 (manufacturing method)	C — Cast F — Forged M — Machined O — Other method
6 (inventory class)	A — A class B — B class C — C class
7 – 8 (material)	Me — Metal Pl — Plastic Gr — Graphite Ot — Other

We could carry on with this part-numbering scheme; however, we now have enough attributes to assign a part number to this part. For this situation, we would assign the number of 31EnCAMe to the part. We have captured all of the relevant traits of the part in this part number.

Several things make this approach very appealing. First, we can now search part numbers for specific traits. We know that the part number consists of segments that describe key traits, and by basing a search on these traits we can impose some form of control over the parts. In the late 1970s and early 1980s, when many North American managers were first introduced to the concept of Group Technology, they became aware of a very disturbing statistic. They were told that between 20 to 30% of all part numbers found in the typical American manufacturing system were for either identical or nearly identical parts. This represented waste, as engineers were basically reinventing the wheel. With significant part numbering, though, managers now have the ability to search part databases to identify items that are identical or nearly identical to the ones that they are contemplating designing from scratch. In addition, by searching for similar parts, an engineer or a manager can draw on past experiences and lessons learned when setting up routings or establishing cost standards for a new part or design. All we have to do is to look in our database of parts to see if we can identify similar parts, and by doing so we can review the routings and costs and time standards. Finally, such an approach is logically more satisfying. We are no longer randomly assigning part numbers; rather, we are using a system to build them so that they reflect key traits of each part.

In spite of these very attractive advantages, the use of significant part-numbering schemes is, for the most part, fatally flawed. In general, such part numbers work very well if the world in which we work is stable; however, in a dynamic world, certain factors work against such a scheme. First, there is the problem of how to handle changes. Suppose that we are currently building a part internally, but, as a result of a "make/buy" decision, it is decided that this part will be outsourced. At this point, we are faced with a critical question — has the part really changed? For the most part, it has not. The only thing that has changed is its source. Yet, to maintain the integrity of the part-numbering scheme, we would have to renumber the part. This would create a potential maintenance problem, as the computer system would have to recognize that the two part numbers (the one for the internally manufactured part and the one for the outsourced part) are for identical and substitutable parts. There is an urge, under these conditions, not to change the part number, but if we do not update the part number then we can potentially jeopardize the integrity of the entire part-numbering scheme.

What happens if there are one or more fields that are no longer needed? If we drop these fields out of the part-numbering scheme, then we run the risk of having numbers that are always changing. For most users who try to remember the part numbers for certain critical parts, this means that they are faced with one of two scenarios. The first is that they are continuously having to learn new part numbers, which does not make for happy workers. Alternatively, they remember only one set of part numbers. The result is that we have a system where different part numbers (all applied to the same part) are floating about the system. This leads to confusion and frustration.

Then there is the problem of dealing with the need to add new key traits. For example, suppose that we are interested in identifying when a specific part is made by a supplier that is ISO/QS 9000 certified. We are faced with the problem of whether we need to add this new information to the part number in the form of a new field. If we allow this change, we run into the potential problem of part numbers that are always growing. One of the authors saw an instance where a firm had a part number consisting of 34 significant digits (and growing). To many of the users, this part-numbering scheme had lost any degree of usefulness. It was becoming too large and unwieldy.

Under such conditions, we have to face the reality that the software packages we are using can no longer accommodate such a large number of digits. What happens if our part numbers are 34 digits, but the software package that we have identified as being almost ideal for our use can only handle 24 digits? In this case, our options are limited. We could pay to have the package modified, which means that we will never have the most current version of the package (remember that every new version must be modified before you can use it). We could look for another package. Or, we could change the part-numbering scheme and generate new part numbers that have no more than 24 digits. Changing part numbers is not something most managers want to do. Many would rather undergo root canals on every tooth before they would even consider such an option.

We are faced with an interesting problem, then. Significant part numbering is attractive because it offers managers several important advantages, but we must weigh several equally important disadvantages against these advantages. We are left with the challenge of resolving this conflict. There is, however, a relatively simple means of achieving this resolution which we will discuss below.

We have discovered in the course of our work what Vince Lombardi, the famous coach of the Green Bay Packers, learned so long ago — it is in the flawless execution of the basics that we find world class success. May you, too, learn and appreciate that lesson.

Part Numbering: What To Do

Overview

Previously, we had examined the pros and cons of a significant part-numbering system. This discussion left us faced with an interesting problem — if there are problems with the use of significant part numbering, then what is the alternative? That is the question that we will address here.

Part Numbering vs. Storage and Retrieval of Information

Far too often, managers confuse two objectives. The first is the use of part numbering as a means of uniquely identifying a part which is the fundamental objective of any part-numbering system; no part-numbering system can be considered minimally acceptable unless it is able to perform this simple function. Then there is the second function of information storage and retrieval. When we talk about significant part-numbering schemes, we are really focusing on this objective.

Storage and retrieval are critical activities that rely upon the dynamics of part numbering. Storage and retrieval require that we address the problem of how to deal with attributes such as whether the part is manufactured internally or purchased from a supplier. We also must deal with the dimensions of the product, its material composition, and its ABC class. As we pointed out earlier, these traits can and do change, so how do we facilitate this aspect of part management? The answer can be found in the *item master*.

The item master can be found in most Material Requirements Planning (MRP)/MRPII/ERP systems and is a record or file uniquely associated with every item or part number. The item master provides descriptive information about the part. Typically, an item master will contain a verbal description of the part which includes such information as the unit of measure (UOM), source of the part (manufacturing or purchased), cost per unit, and inventory class. What is interesting about this record is that in many systems (especially those that are based on one of the major database languages) the item master can be modified as needed. That is, if we need to make available new items of information, then we can add new fields to the item master; however, if we find that these fields and the information that they contain are no longer relevant, we can just as easily drop them. For example, suppose that we were interested in being able to determine if the supplier of a specific part was Y2K compliant. We could have modified the item master to include a field that indicated whether each supplier was Y2K compliant. This "Yes/No/Pending" field (allowable values that we would find in the field) could have been kept and searched until January 1, 2000. At that point, it would no longer be needed and should have been dropped. By incorporating such a field into the item master, and not in the part number, we can minimize potential problems. We still have access to the necessary information; however, in this case, there was no need to change the part number to reflect whether the supplier was Y2K compliant.

The item master gives us access to a record that is sufficiently plastic that it can be changed to meet our current needs and requirements. That is not all that the item master gives us access to, though. The item master also facilitates retrieval of information because, in most systems, the item master can be searched. Further, because each feature of the various parts is stored in its own field, we can conduct targeted searches. If we are interested in identifying those items that are A class, all we have to do is to search the field that contains the inventory class assigned to the parts. We do not need to break apart part numbers and focus in on those locations that contain the inventory class. The result is a very efficient and effective search.

In addition, the item master gives us a simple method of dealing with the problem of how to manage multiple part numbers assigned to a given part. On occasion, we might encounter a part that is described using different part numbers. One of the authors encountered this problem when talking with one of his past students. This student, who works with one of the automotive firms, can best be described as really sharp and top notch. He told the author that, in his operation, there were parts that were described by as many as four part numbers — an internal part number, a service part number, an old part number (no longer widely used but still remembered by many of the managers within the operation), and a part number that came from overseas (the operation was initially a joint venture). At any point in time, he was faced with the problem of determining which part number to use. With an item master, these various part numbers could be included in their own field and immediately cross-referenced to the basic part number. No matter what part number we might be given, we can enter it into the search to find the underlying part numbers. Also note that this cross-referencing can be made transparent through the use of macros or programmed front-end.

In short, it is the item master that allows us to work with a wide variety of part numbers. At this point, a number of issues still must be considered, such as what we can do if we do not have access to an item master of this type. In this case, there are only a couple options available. The first is to add this capability to our software, which can be done by the software vendor or by buying a part-numbering/item master package that works with our software. The second option is to change software packages altogether.

For others, there is the issue of what to do with the part numbers if they reflect a mess of different approaches. In this case, the first reaction might be to take the position of not worrying about it as long as these part numbers are truly unique. If they truly are unique but we are not satisfied with the lack of consistency in the part-numbering schemes, then we should recognize that bringing about this desired consistency is not really required to make the part numbers work; we are simply cleaning up the part numbers, which is a major undertaking and one that we have to consider very carefully in light of the relative costs and benefits.

Comments from Column Readers

When we began writing our APICS column on the topic of part numbering, we never expected the level of interest and feedback that resulted. It appears that part numbering is something that most people see and use but few understand. The following comments are typical of ones we have received from our readers:

- There is no such thing as a part number that can be "not significant." It is better to talk in terms of descriptive or nondescriptive.
- While part numbers can consist of both letters and numbers, it ultimately makes sense to have part numbers that are all numeric. Alpha causes all kinds of problems. Why do you think that the telephone numbers are now only numbers? They used to have exchanges (here the reader is revealing his true age).
- Engineering change management is an important element of part numbering management; however, it is too important to be left to engineers.

Part Numbering: When To Change the Number

So far in this series of discussions on part numbering, we have addressed a number of critical issues; however, there are two important questions that have yet to be addressed:

1. Do prototypes require a different part number?
2. How do we handle design revision?

Both of these questions are interesting in that they really deal with one central issue — under what conditions do we change a part numbering? Because this is a critical question, it will form the focus of this discussion.

How To Handle Prototypes

Prototypes are a fact of manufacturing life. It is in the prototype that we evaluate the functionality of the proposed part; it is also in the prototype that we assess traits such as ease of manufacturing. Because prototypes are "tentative" new parts (tentative in that they are subject to change based on feedback received from the shop floor, marketing, users, or engineering), we would like to inform the people working with the part of its tentative nature. As a result, some firms and managers have adopted the practice of assigning these parts their own unique part number. Once the part has been debugged, the part is then given a different part number to indicate the change in its status.

On the surface, this practice makes sense. It helps to uniquely identify a part and its status, and it conveys some important information about the part. However, this practice should, if at all possible, be avoided because it is wasteful in that it consumes part numbers. Consider the following situation for a moment — you have decided to use an 8-digit part-numbering system, which means that you can assign up to 99,999,999 different part numbers. Potentially, though, the practice of using prototype numbers can result in your losing half of your part numbers. How? Because every part in the database can have two different part numbers — a prototype number and its ultimate number. If this is the case, then what can we do? The answer is fairly straightforward.

Previously, we had discussed the importance of the item master/part master, with its searchable user-defined fields. The key to handling prototypes is to turn to these user-defined fields. Remember, the purpose of giving a prototype its own part number is to uniquely identify it for the user. This same outcome can be easily achieved by defining a new field: part status. This field is alphanumeric in structure and consists of a number of possible values. One of these values can be "P," which denotes that the part is a prototype. With this piece of information, you can now define a new part number that consists of the ultimate 8-digit part number along with a one character suffix. For example, we could denote a part with an ultimate part number of 12345678 as a prototype by adding P to it. The result would be the part number 12345678P. When the part is finally debugged and ready for production, we can drop the suffix P, leaving us with the ultimate part number. The advantage of this approach is that it does not affect those structures and database items dependent on part numbering — structures such as the bills of materials.

This suffix approach is also highly appropriate for dealing with situations such as flagging a part that is being considered for deletion or flagging a part that we want people on the shop floor to study. For example, this is a nice way of implementing "red tagging" within the computer database.

This approach really forces us, as production and inventory control managers, to recognize that part numbers are a corporate asset and resource. They should be managed as carefully as inventory, money, or people's time. This is a view that very few firms and managers have adopted.

How To Handle Design Revisions

It is a fact of life that ultimately every part or item that we design or build will be eventually modified. These modifications occur for a number of reasons — changes in materials, changes in processing technology, product improvements/enhancements, changes in the parent products, or phaseout of the parent product. Every time we carry out an engineering design, we are changing the part. This leads to the ultimate question of determining under what conditions these changes are sufficient to cause us to assign a new part number to the affected part.

The answer to this question is both simple and complex. In short, we should assign a new part number whenever the part has been fundamentally altered. Now comes the challenge — determining when the part has been fundamentally changed. In short, we deem a part to be fundamentally changed when it no longer fits the previous needs. These needs are found in both the function and processing. On the first dimension, we must change the part number if the resulting changes effectively prevent the "new" part from being substituted for parts made over a period of time, such as the last 5 or 10 years. That is, if you cannot take the redesigned part and use it without modification to substitute for the same part as designed and manufactured 10 years ago, then you must change the part number. Why? Because the design changes have affected our ability to service our customers.

On the second dimension, the part number must be changed whenever the design changes affect the processing requirements sufficiently so that the processing times or incurred costs are now different. Before going any further, it is important to recognize that management must identify the conditions in advance that define when a new part number is needed. These conditions must be identified in advance so that we can prevent people from applying their own "rules," a situation that ultimately results in increased confusion as some parts receive a new part number for changes that seem not to affect the part numbering of other parts. Further, all proposed changes in part numbering should be reviewed by management to ensure that the rules and procedures have been applied consistently.

With this note, we have finished our examination of part numbering. As can be seen, part numbering, while basic to every firm, is still one of those basics that, in general, is very poorly understood.

Part Numbering Revisited

Overview

Of all the topics discussed in our APICS column, this one has generated the greatest amount of discussion. It seems that someone is calling us almost every day with questions about part numbering. Further comments we have received from our readers can be summarized as follows:

- It is important to remember that part numbers are used by people. As we add digits to the part number, we are not simply increasing the length of the part number but also increasing the probability that an error in recording or transcribing the part number will be made. The magic limit for the number of digits in a part number is eight. As we go beyond that, we increase the probability for errors.
- The problem with significant part-numbering systems is that we can never be sufficiently comprehensive. A new requirement or trait will always show up that should have been treated adequately in the past but now poses a problem for us. As a result, users may compromise on the part-numbering logic, and the integrity of the part-numbering system is compromised. As soon as this happens, the usefulness of the significant part numbering structure is adversely affected.
- Why are we worried about the length of the part number? After all, we know of packages that can store up to 70 significant digits.

Another Look at Part Numbering: One Reader's Perspective*

In a December 1998 article in the APICS journal, Melnyk and Christensen made a strong case for "dumb" sequential part numbers. I agree with them and want to amplify this discussion.

Modern Part Number Usage: *In today's computer-oriented information world, part numbers are no longer "identifiers" but "differentiators;" that is, they are primarily computer tools rather than people tools. The part number establishes the uniqueness of each item, but we have to look beyond the part number itself to determine the "unique" properties of each part. What if we want to order the new APICS Dictionary, 9th edition? We would go to the 1999 Educational Materials Catalog and look in the index for the APICS Dictionary, and are then directed to go to page 21, where we find the item in question, a short description, the price, and the stock number (#01102). We can go to the APICS home page, use FAXBACK, or call APICS headquarters to order the new dictionary without any knowledge of its stock number (a.k.a.*

---

* By Don Frank, a well-known production and inventory control consultant.

part number). If someone had ordered stock #01102 last year, the 8th edition would have been shipped. Order today, and the 9th edition will be shipped. The part number has not changed, but its "revision status" has. The management of part revision without changing the part number is a good subject for another article.

Part numbers should be viewed merely as entrance keys to all the data peculiar to a given part. With today's sophisticated search engines associated with part databases, the focus should move to the description as the primary tool for locating unknown parts. One can, with proper use of the primary description field, search for a screw that is 10-32, 1.25-in long, steel cad-plated, and cross-recessed and then display data about all the parts that meet the search criteria. The lesson that should be learned from this is that it is our descriptions that should be cleaned up and standardized before we undertake changing part-numbering rules.

Part Number Ownership: *Part numbers should be viewed as enterprise tools, rather than being owned by one particular function or department in the company. In the global enterprise requirements management (GERM) world in which we now live, part numbering has ripple effects down to suppliers' suppliers and up to customers' customers. Therefore, we need to take a much wider view of parts and part numbering than we have heretofore.*

In many companies today, part generation is totally balkanized. This process has only been exacerbated by the emergence of component source (supplier) management (CSM) tools. We now have internally generated parts controlled by the CAD/CAM process, with their set of rules; component parts controlled by the components engineering group, with their set of rules; and, finally, the know-it-alls of ERP implementation, who create their own set of rules for the BoM parts in MRP (with or without the advice and consent of engineering or procurement).

Enlightened companies will use the engineering drawing number as the part number for traditional parts as well as specifications, test procedures, and other documents that should be listed on the bills of materials. The use of tabulated drawing/part numbers has the advantage of grouping all like parts onto one specification control drawing for standardization, eliminating a separate field for drawing number and simplifying the process of part configuration management. However, one must remember that prefixes and suffixes are not appended to part numbers; they create new part numbers.

Changing Part Numbers: *One of the real concerns today, as was pointed out previously, is the duplication of part numbers for the same basic part and the number of "look-alikes" that could perform the same function. The tracking down of these meanies should be the function of the cross-functional part standardization team, and should be an action item in all implementation or system upgrade projects. The implementation or upgrade process presents a* *wonderful opportunity to identify and obsolete the duplicates and look-alikes. Notice that the term used was obsolete, not delete. The old part numbers should be retained and coded as obsolete, and their descriptions should read: "Obsolete, see part [new part number]." This protects against having to deal with a customer replenishment order where only the old part number is in the customer's records. Having done the standardization, all parts that truly have the attributes of the new, standard parts should be placed in the same bin as the new parts; parts that do not meet these criteria should be scrapped and destroyed. Closing the loop, an online or printed "Standard Parts Selection Guide" will help to reduce future part duplications.*

Readers should be cautioned against arbitrarily retiring duplicate parts and/ or going to a new part-numbering scheme. There is so much history in drawings, cost records, customer provisioning and maintenance files, etc. that one cannot afford to lose access to these data. So, what to do? The technique of standardizing is only part of the answer. Consider the situation many companies find themselves in today, where a merger has combined six product lines from three companies, each with their own part-numbering scheme(s). 'Tis utter chaos! One enlightened solution is to have alternate keys in the database to [new_part_number] and [old_part_number] so that folks can have their cake and eat it, too. This approach is also recommended when going to new, standard, sequentially generated part numbers. Anyone who suggests a new part-numbering scheme without retaining an easy path to the old part numbers should be drawn, quartered, and hung out to dry.

Conclusion: *In my 50 odd (and even) years in this business, the only path out of this wilderness that I have been able to use effectively is to create an Enterprise Part Czar, providing one control point for the generation and maintenance of part numbers and their associated data. Making the generation of new parts a somewhat cumbersome process is not all that bad an idea, since it can get the lazies out there to research existing parts before creating look-alike or duplicate new ones.*

Part numbers and part numbering have come a long way since we were forced to generate them in the old punched-card days. We should obsolete some of that punched-card, batch, mainframe thinking and start to use part numbers and part numbering in an up-to-date, efficient, and user-friendly manner.

Major Lessons Learned

- Ultimately, it is important to remember that part numbers are used by people. As a result, we must note that adding digits often increases the opportunities for error.
- We can use a significant part-numbering system, provided we are sure that we can develop a scheme that is sufficiently comprehensive — a major challenge.

- Part numbers should facilitate our ability to search the computer database for locating specific parts.
- The issue of ownership of the part number must be explicitly addressed. Without ownership, no one is responsible for part numbers — a sure recipe for the deteriorating effectiveness of the part number.
- Obsolete part numbers should be carefully managed within the firm.

Bills of Materials: An Introduction

The Problem

Some six years ago, a furniture company located in western Michigan contacted us and asked us to help in structuring a training program for the company. In the process of assessing operations at the plant, we had a chance to talk with some of the engineers. The engineers were proud of their bills of materials. They noted that their bills were, at a minimum, 95% accurate. No other firm in their industry, they pointed out, could even hope to approach this level of accuracy.

After talking with several other people in the firm, though, we heard that many of the more commonly used C items were purposely omitted from the bills. As another person noted, "The bills are 95% accurate, but only 75% complete." When this observation was shared with the engineers, their reaction was one of "So what?" They knew that they had not included all of the items in the bills, but the items that were not included were very common and frequently used. As an example, they pointed to the wooden corner blocks. These blocks were used in every desk, table, and chest. These were C items by any definition of the term. Currently, these items were managed using a grocery store approach. Bins of the corner blocks were located on the shop floor. Each bin had a line inside it. When the parts fell below the line, it was the responsibility of the parts manager to place an order for replacements. Because of the nature of these parts, it was argued, there was no need to include these items, thus needlessly cluttering up the bill of materials. It was at this point that we came to the realization that these engineers had confused the role of the bill of materials with inventory control.

The problem with the logic of these engineers was that they failed to understand the purpose and role of the bill. The equipment that was used to cut the corner blocks was also used for cutting other wood. Even if the corner blocks were cut at off-peak periods, these blocks still demanded capacity. As a result, the capacity implications of the corner blocks had to be considered. Bills of materials facilitate these types of calculations. This lack of understanding is not isolated to this company. As we have worked with other companies, we have noticed the same lack of understanding, so it is about time to clarify the concept of a bill of materials.

Defining the Bill of Materials

According to the APICS Dictionary, the bill of materials is defined as:

> *...a listing of all the subassemblies, intermediates, parts, and raw materials that go into a parent assembly showing the quantity of each required to make an assembly. It is used in conjunction with the master production schedule to determine the items for which purchase requisitions and production orders must be released.*

This definition touches upon one of the major purposes of a bill of materials (BoM). A BoM is essentially a recipe that helps to answer the questions of:

- *What:* What components/parts/assemblies are needed to complete one unit of an end item? The identification of these items is achieved using the part number.
- *How much:* How many units (depending on the unit of measure) are needed to make one unit of the end item?
- *In what order:* The bill of materials identifies the exact sequence in which the parts, raw materials, and such become components, assemblies, and, finally, the parent items. The sequence is revealed by observing the manner in which the parts are organized by levels.
- *When:* This question addresses the issue of time. In the bill of materials, we must deal with planning lead times, as these determine when a component must be ordered so that it arrives when it is needed.

A BoM is necessary for formal planning, both material planning and capacity planning. We cannot do any form of material planning (whether it is through Material Requirements Planning or not) without some form of a bill of materials. Further, we need bills for system control. They are used for setting effectivity dates, for dealing with the phase-in/phase-out of end items and parent products. That is, when assessing when to phase-out a product, we often look at the component inventories currently available for the product. We use the structure provided by the BoM to determine when we can expect to run out of these parts, and we can use this information to determine when to phase-out the parent and when to phase-in its replacement. We use bills for pegging; that is, when we encounter a problem with a component, we use the bill to identify the parent items that could be adversely affected.

Finally, we use bills to facilitate capacity planning. We often tend to overlook this function of the BoM, but it plays a critical role. When we explode an end item (i.e., convert the requirements of the end item into its component requirements, after adjusting the requirements for the availability of inventory and on-order parts), not only do we identify the components needed, but we also generate demands on capacity. Identifying these capacity requirements is done by adding to the bill of materials new information through the bill of resources. When we focus on capacity, we are interested in determining if we have sufficient capacity (as well as material) to meet the needs of the parent items. It is because of this factor that we cannot accept the argument that an acceptable bill of materials can be "95% accurate and 75% complete."

Then what we should expect a bill of materials to be? We want an acceptable bill to be the following:

- *Accurate.* It should include all of the components (in their correct order) necessary to build one unit of the end item.
- *Current.* It should represent how we now build one unit of the parent item, not how we used to build it.
- *Complete.* It should include all of the components needed to make one unit of the parent item, with no exceptions. If we omit one component, we tend to ignore it. We can only plan for it by relying on the memories of the people involved. In other words, when we omit a component, we implicitly rely on the informal system to cover for this omission.

Bills of Materials: Understanding the Uses

Overview

We have shown that many users have an incomplete understanding of this key element of every effective and efficient production and inventory control system, and we will now explore the various uses of the bill of materials. By focusing on these uses, we reinforce the concept that the bill is something that should not be taken lightly.

Uses of the Bill of Materials

Bills of materials are important because, plainly put, they play so many important roles. Put in another way, we need good bills because we need the functions that they provide. The poorer the bill, the less able it is to fulfill these functions. Following are some of the more important uses of a bill of materials.

Product Definition

The bill provides critical information on how to build a product. It is the equivalent of a recipe. It tells us the items needed, in what quantity they are needed, and when they are needed. It also helps to distinguish between those items that are built inside the plant and those items that are acquired/supplied by our vendors. As a result, such information helps the user to identify those items that require the intervention of the purchasing department and those items that could be managed through the shop floor control system.

Engineering Change Control

One of the objectives of any bill of materials system is to ensure that the bills are always completed, current, and *accurate*. This is critical because of the bill's role in defining the product. When viewed from this perspective, we can see that there are many ways in which a bill can be used. It can be used to facilitate engineering change control. When a potential change in a product is introduced, we can use the bill to identify those items and components potentially affected by this engineering change. These could be parents that use the affected part, or they could be components used in making the affected part. For these latter components, we might be interested in determining if they are used in any other parent. If they are not, then we have identified a "hidden" cost of the engineering change. Typically, when we introduce engineering changes, we focus on the specific part. We rarely consider the impact on those components used by the part. These considerations can influence the "effectivity date" — the day when the engineering changes are introduced or turned on.

In addition, the bill of materials, especially when displayed in a graphic form, can help identify opportunities for product simplification or component standardization. That is, we can look for unique components — components that are specific to one and only one parent. The costs of maintaining such a unique part are often greater than any benefits. These items become candidates for elimination or substitution by a more standard component.

Service Part Support

Because the bill defines the product, it is often used by the service department to identify what components they need to order when a customer calls in to complain about a product breakdown. If this bill is wrong, then the service department will inevitably ship out the "wrong" part. This is not a malicious action; rather, the service person is simply basing their actions on the information provided in the bill. The problems created by poor bills were driven home to us when we visited a plant located in the Midwest.

This plant built RV chassis and other forms of specialized chassis. Many of these chassis were highly engineered. A problem that plagued this firm was that, because of tight due-date requirements, changes were often introduced into the chassis with the expectation that the bills of materials that accompanied these vehicles would be eventually updated later; however, this seldom, if ever, occurred. As soon as the management and the shop floor personnel had finished putting out one fire, they were faced with another crisis demanding immediate action. We saw one instance when the shop floor people were jury-rigging a pedestal for the driver's side seat while the engineering personnel were running back and forth with engineering changes. The last thing on anyone's mind was keeping the bill of materials complete, current, and accurate.

Over time, when these vehicles had gotten out into the field, they began to wear down and require maintenance. The owners, who had typically paid over $250,000 for a finished RV, could find themselves stranded on the road. When they were able to call into the firm to order replacement parts, the service department personnel would then refer to the bills of materials to send out the necessary replacement parts. Inevitably, when these parts arrived, the mechanics found out that they were the wrong ones. Now everyone was faced with a really nasty situation. There was the service department, which did not know what types of parts to send out. There was the owner, who had paid a great deal of money only to find that a broken-down RV has very little value. Then there was the mechanic, who was caught in the middle of trying to get the correct replacement parts and calming down a very irritated customer.

To address this problem, the manager of the firm had introduced a program that promised RV owners that they would be up and running within 24 hours, or the firm would pay for their lodging or travel expenses. Because nothing had been done about the underlying cause of the problem (inaccurate bills), the result was that this program generated a significant drain on the cash reserves while doing little to pacify the angry customer. In most cases, the firm would send out a mechanic from the plant (armed with a selection of possible parts) to correct the problem. Eventually, everyone agreed that the money spent on this customer program would have been better spent on making the bills accurate.

Warranty Analysis/Control

Every product is uniquely defined by its bill at the moment that it is built. As revisions are introduced, the product is changed and these changes are reflected in the bills. When warranty problems are encountered, the firm can use information about the product at the time it was built to identify potential causes of these warranty problems. This information can be used in evaluation procedures such as failure modes and effect analysis (FMEA) to determine what specific features of the product are most likely responsible for the problems. This information, in turn, can trigger subsequent changes in the product.

Planning and Scheduling

Because the bill tells us the order in which the various components are needed (information found in the lead-time estimates), we can use this information to determine how to schedule components supplied by our plants and suppliers.

Costing

The bill identifies the components needed. This information, in turn, can be used to determine the costs of the end product. We can begin from the bottom of the bill and roll up the costs by adding up the costs associated with each component.

Shortage Assessment

To this point, we have worked from the end product back towards the individual components; however, we must recognize that there are many instances in which a component becomes short (a vendor experiences a problem, a machine breaks down). At this point, we have to determine what parts are going to be influenced (and how). This assessment is done using the bill of materials.

Backflushing

In many high-volume environments, we determine what components have been consumed in production by means of backflushing. That is, we have identified certain checkpoints in the process. Whenever an order passes such a point, we use the bill to identify the components that should have been used. With this information, we can get an idea of the accuracy of our inventory records (actual ending inventories on hand should equal the beginning inventories plus receipts minus backflushed withdrawals). Backflushing is not feasible without accurate bills.

Terms and Definitions

As we can see, the simple bill of materials really satisfies a number of very important functions and uses. It is central to the operation of every product and inventory control system. We will now continue our discussion of bills of materials by looking at various terms and definitions closely associated with them.

If the e-mails and telephone calls received by the authors are any indications, then our columns on bills of materials have generated a great deal of interest. This interest only serves to reinforce the premise that the basics, while critical, are often poorly understood. One of the people who communicated their concerns to us was Don Frank. If any of you are readers of the APICS journal or have followed the APICS proceedings, then you are familiar with Don Frank and his contributions to the field. Don is someone who can best be described as a seasoned, highly knowledgeable manufacturing silverback. After reading his e-mail to us, we invited him to write down his ideas about bills of materials. The following is his response. We hope that you find it as enlightening as we did.

Revisiting Bills of Material[*]

We will look at the bill of materials data elements, their creation, functions, and maintenance. When we view BoMs, either on screens or hardcopy, we are looking at data from both the BoM and item master files. BoM data, whether in flat files, relational databases, or objects, are stored as parent-component pairs. It is the bill of materials processor that reaches out to the part file for such data elements as part description, unit of measure, revision code, etc. The basic elements of the BoM file are parent part number, component part number, quantity, yield factor, and date effectivity. It is the bill of materials processor that collects these parent-component pairs into single-level, indented, or summarized bills of materials by walking down the product structure trees, or into single-level, indented, or summarized lists, where used, by walking up the trees.

The quantity in the BoM must be consistent with the item master unit of measure. The yield factor in the BoM must be carefully considered and not abused because the item master also contains the safety stock quantity and shrinkage factor. One must decide how to use these three data elements so they do not compound and create more inventory than is intended or needed.

Most BoM systems, unless otherwise directed, will present the BoM in part-number order; however, another BoM data element that is often used is the "find number" (also known as "balloon/bubble number" or "reference designator"). The bill of materials was originally on the assembly drawing, along with the find or balloon numbers associated with that BoM line item with its place in the drawing. In electronics, this concept has been expanded to include schematic diagram reference designators, which associate each component on the schematic with its location on the assembly drawing. This application of BoM data is very useful in relating physical parts to schematic diagrams or discussions, for automatic part insertion in printed wiring boards, or for their function in instruction manuals. However, because the part number for many different components may be the same (R1, R4, and R17 can all be part 12345), it is necessary for the bill of materials file to allow more than one "hit" on a part number in all single-level BoMs and for MRP to accumulate the usage when doing its gross-to-net calculations.

Perhaps the most misunderstood BoM data element is date effectivity. If the field has a single date, the message is "do not use this BoM line before this date." If a second date is used, the message is "do not use after this date." The net result is to establish a window so that MRP will not use this BoM line outside of the effectivity date. Its use is to phase-in and phase-out lines in the BoM based on engineering change data. There are a number of problems with this concept.

First, is it realistic to assume, in today's fast-paced enterprise planning and execution, that a date can be established 4, 6, or even 10 weeks into the future

---

[*] This section was written by Don Frank.

and that the date will remain valid? If one is to establish and maintain credibility for the BoMs, these dates must be constantly reviewed as the planning process changes.

Second, because the bulk of the changes come from engineering, one must recognize that engineers normally think in terms of revision codes (going from Rev. A to Rev. B) and are not often charged with the responsibility of engineering change execution by determining the best implementation strategy to minimize disruption of the manufacturing schedule and to reduce the obsolescence of the parts being phased out.

This paradox can be resolved only if the engineering change process is a team activity with both engineering and planning people involved in the effectivity process. Two lessons learned from concurrent engineering are applicable here:

1. Having manufacturing and planning people involved in the design process significantly reduces the number of engineering changes.
2. Using the same team approach to planning and executing engineering changes makes the date effectivity process more reliable.

Other ways to set effectivity that may be more reliable than the classic date effectivity include:

- Work order effectivity, where the effectivity is set by individual work or shop orders
- Lot effectivity, where the changes are applied to specific manufacturing lot numbers, either indicating the first lot affected or specific lots where the change applies
- Serial effectivity, where the effectivity is applied to each individual unit made and shipped (used extensively in the aircraft industry where each "tail" has its own effectivity)
- Customer order or contract number effectivity, where effectivity is defined to particular customer deliverables

The best rule is to choose the effectivity tools that best meet the enterprise business process and then apply them using cross-functional teams.

Another concern should be the time it takes to process engineering changes. Product Data Management (PDM) and its workflow tools have allowed the circulation and approval of engineering changes to be done electronically, rather than by paper circulation. While this is a step forward, it does not guarantee that the right information will get to the right people quickly. One good idea is to use Pareto techniques to assign A/B/C ratings to changes:

- An "A" change is a critical one for which preliminary action must take place immediately. Even before the change is approved, planners and foremen are authorized to stop work on items subject to an "A" change so as not to buy, issue, or make parts that will have to be reworked once the change is approved. The rule is that the sun will not set before preliminary action is taken.

- A "B" change is one that can be scheduled based on the run-out of the parts to be replaced and/or the best time to insert the change in the shop schedule. Effectivity setting should be done within 5 working days.
- A "C" change is either a minor cost improvement or data change that does not affect the form, fit, function, or fee (the 4F rule) of products. These changes should be accumulated and processed as a group either when product upgrades are scheduled or in quarterly cleanups.

In conclusion, we would do well to remember that bills of materials, while created and maintained by engineering, are enterprise tools and that all stakeholders have a vested interest in their accuracy and the timeliness of configuration changes.

Major Lessons Learned

- The most misunderstood BoM data element is date effectivity. This is used to control the timing at which one BoM is made current and another is made obsolete.
- Managing date effectivity requires a team approach, as it involves not only production and inventory control but also engineering.
- When managing the phase-in of BoM changes, we should use an A/B/C approach, where "A" changes are implemented immediately, while "C" changes are periodically implemented (typically when it is convenient).
- Bills of materials should not be viewed as production and inventory control tools, but rather as enterprise responsibilities and assets.

10 Getting Things Done, and Done Right, the First Time: The Basics of Implementation

Facta, non verba ("actions, not words")

Implementing the Savings: The Dumb Way of Achieving Managerial Excellence

Overview

No discussion of the process of generating savings would be complete without a discussion of how we go about identifying savings potential and achieving savings. Every organization goes through the steps every year to plan for changes that will generate savings through the annual goal and objective planning process, but that is just where we fail in our attempts to generate savings. We look at the savings that we think we can get and set our goals on achieving those savings. There are a lot of business books that are written on just that premise: Identify the savings potential and implement. That is the problem, though. We tend to look at the savings that can be achieved without ever defining the process that needs to be implemented to generate the anticipated savings.

What Are the Savings?

When we look at the goals and objectives of our corporations, we sometimes get the feeling that those goals really are the anticipated savings. Most companies have a goal that might sound something like this: "It is our goal to reduce the work-in-process inventory by 10% this year." A lofty goal but one that most companies set for themselves. Then we spend the rest of the year trying to attain that goal and usually fail because we do not know how to achieve the savings. We have a really easy way to achieve this goal — stop buying materials for about a month, which should pretty much achieve the goal of a 10% inventory reduction. We have accomplished the goal that the company established and have met the objectives and needs of the company, but this is the wrong solution to the problem.

This example, while obviously not being a valid solution to the problem, does illustrate the problem quite well. What we really did was to implement the savings. The goal was to reduce inventory, and that is just what we did. What we did not do was to implement the tools necessary for us to reduce inventory the proper way. We selected the wrong tools to generate the savings in inventory. There is no doubt that there really was a reduction in the inventory level, but the corporation is not able to sustain operations at that lower inventory level based on the method that we used to generate that reduction. This was a case of implementing the savings, which is a not a way to achieve excellence in our operation.

We need to implement the tools that will allow us to run with the lower level of inventory. When those tools are in place, only then can we lower the inventory levels down to a point where we can now safely operate at the new lower inventory levels based on the tools that we just implemented.

Here is another example. A few years back we had a person, who will remain nameless (and we do apologize if you are reading this and recognize yourself, but you did earn your place in this book), approach us to say that, "This smaller inventory lot size concept doesn't work. We tried it and it is no good." We asked him how long it took before he went on backorder, and he gave us that blank look of amazement and said, "We went on backorder in 6 weeks. How did you know?" How did we know? It was easy. That company had done what this section is all about. They implemented the savings. They were trying to achieve managerial excellence but chose to implement the savings. They saw the goal of smaller lot sizes as a solution to a problem, but rather than implement a strategy that would allow them to reduce inventory they went straight to the savings they were after (smaller lot sizes) and implemented smaller lots by forcing the issue. They just changed the size of the production order one Monday morning and thought they were done. They implemented the savings. There really are savings to be gained by working with smaller lot sizes, but just cutting the lot size in half is about the same as implementing a 10% inventory reduction by just not buying anything. That is not a tool that will get us the savings we need.

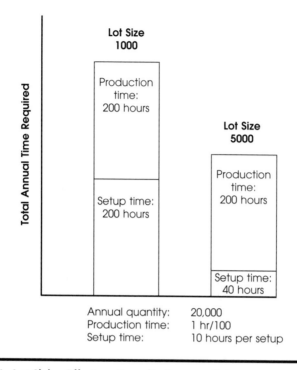

Annual quantity: 20,000
Production time: 1 hr/100
Setup time: 10 hours per setup

FIGURE 16. Lot-Sizing Effect on Capacity Consumption

We need to look at what really happened when the corporation cut the lot size in half. By doing so, they effectively doubled the number of setups required to achieve the same annual production. Suppose that, in this example (as summarized in Figure 16), the corporation sold 20,000 units of a particular product per year, and they had a lot size of 5000, thus requiring four production setups per year to build the 20,000 units. Now they cut the lot size by 80%. Instead of four production runs per year to make the 20,000 units, they now require 20 production runs per year, at 1000 units per production lot, to achieve the same results. In effect, they have quintupled the amount of setups required and quintupled the amount of setup time required per year to get the same amount of production. If the setup time was 10 hours per setup, under the old system they would have used 40 hours to set up the four production runs. But now, with the smaller lot sizes, they need 200 hours per year. Where did that additional setup time come from? It came from run capacity, or the time that they needed for production.

If we have a factory or piece of equipment that is not loaded to 100% capacity, then the additional time required for the setups came from the idle machine time. If, however, the factory or piece of equipment was loaded to 100%, there is only one place that the additional time for the setup came from.

The setup time had to come from the production time, and in this case there would be 120 fewer production hours available per year to complete the orders. With plant capacity being used in that manner, and with no additional capacity acquired through overtime, etc., the plant could do nothing other than to go on backorder. That is how we knew that they went on backorder. They used production time they needed for non-value-added, nonproductive setup time. Production capacity had been sacrificed for the additional setup time needed. They had implemented the savings.

What they should have done was to reduce the amount of setup time required first, which entails identifying the tool necessary to meet the desired goal or objective and generate the savings. In our example, a program should have been put in place first to reduce the setup time to a point where it was cut in half or reduced from 10 hours down to 5 hours. After the setup time had been reduced and the processes stabilized, then the plant could cut the lot size in half. Implement the tool, then take the savings. This company would then be stabilized at 24 lots per year at a setup time of 5 hours per lot for an annualized setup time of 120 hours. We need to look closely at the driving function or objective we wish to attain and then determine what needs to be done to achieve savings. Implement the tool, then take the savings. Reduce setup time, then reduce lot size.

We realize that these examples are quite simple and are rather obvious. But are they really that obvious? The lot-size guy in the example shown didn't get it, and it took a lot of discussion before what we were telling him began to register. He could not see the total effect on his organization because he never looked at the entire year to realize how much run time he had lost to setup. He did not look at the total picture to see the effect the change had on his annual production capacity. For the first few weeks, things went great. They always had product on hand and were meeting customers' orders while at the same time their inventory was going down. All their goals were being achieved. But in six weeks, the backorder problem came up. Because of the time lapse between the lot-size decision and the backorder problem, they did not associate the loss of production capacity with the backorder problem. They could not see the problem because of the short-term window that they were looking through. They implemented the savings and now could not determine how to get out of the problem.

We have a short exercise for you. We want you to take a larger view of your operation and look at your own goals and objectives. There is a high probability that you have goals such as "reduce inventory," "shorten lead times," "improve inventory turns," "increase on-time shipping percentages," "improve order fill completion percentages," and on and on. Most companies have such goals and objectives, and many others, that they want to accomplish during the next year.

Take another hard look at these goals, though. What we are really looking at are the savings. These are the dollar savings that management wants to achieve in a relatively short time period. And, if we try to generate these gains from these objectives, we end up implementing the target goals, but what we are really doing is implementing the savings, because nowhere in this list of achievable goals is there a list of the tools that must be put in place first before we can take the savings. As in our example, there are no programs for setup reduction listed in our goals. Nothing lists the tasks that must be completed and in place before savings can be achieved.

Management has a lot of fancy names for what they want to accomplish, calling them goals, objectives, strategies, opportunities, plans, etc., and the list goes on and on. Take a good look here, though. There is no mention of the tools that need to be in place in order to realize the savings.

Major Lessons Learned

This section illustrates that implementing the savings is the dumb way to achieve manufacturing excellence. We need to implement the tools first, then take the savings. There is no other way to do it. Yes, we looked at some simple examples, but they do show what can happen when we begin to implement the savings without having the tools in place — a simple concept that is frequently overlooked.

The Power of WIIFM (What's in It for Me?)

Overview

Be warned. This section deals with something that is based more on management than on production and inventory control. It deals with a basic concept that many managers overlook on a regular basis. This is the concept of WIIFM ("What's in it for me?"). Before discussing this management basic, we should give you some background.

One of the authors was working in his office when he received a telephone call from an individual who presented him with a problem. This person had learned about setup time and the need to reduce setup time. He had also read about the functions of inventory, and he was interested in merging together these two concepts. To that end, he had identified a process, and he had assigned a team to reduce the setup times on that process. They were successful in that they had reduced the setup times from some 14 hours to about 8. Because it did not seem as though the setup times could be reduced any further, the caller had decided that what had to be done now was to introduce a lot-sizing approach that would allow the company to set more economical production quantities. Given the bias against lot sizing that the authors have, this decision was not taken too well.

As the conversation progressed, the caller informed the author that the people involved with the setups viewed them as somewhat interesting. They enjoyed doing setups. In contrast, these same people found production repetitive and boring. After listening to this conversation, the author told the caller that he could not expect any more reductions in setup times. There was nothing in the setup reduction activity that could benefit the people involved. Because they saw no benefit, they would not do it. If they did anything, it would be the minimum amount that they could get away with. In short, the WIIFM was missing. The caller was surprised. Setup times should fall simply because he had told the people that they had to reduce them. The conversation ended in frustration. It was evident that the caller did not understand the power of WIIFM.

The Power of WIIFM

As everyone knows, managers get people to do things; however, how the people assigned carry out these tasks is greatly influenced by the extent to which they see the actions as helping them or hurting them. When the outcomes of the assigned tasks are seen as being consistent with the self-interests of the people involved, these people will work harder at solving the problems or uncovering the desired solutions. When the outcomes are seen as being counter to these interests, they will try to avoid being involved or will do the least amount possible. At times, they will even try to undermine the tasks (a fact that can be attested to by anyone who has ever tried to carry out a time study in a shop where the labor-management climate has not been all that cordial).

"What's in it for me?" is a powerful force and one that every manager should recognize. Before a task is assigned, the manager should take a moment to figure out how this task would benefit the person involved. There are at least five ways to generate such benefits, the first being a promise of *survival*. People will do things if they know that the alternative is the loss of income, hours, jobs, or even overall employment (should the plant be closed). As someone so aptly put it, "There is nothing like a hanging to so focus the mind." For example, returning to the scenario described at the beginning of this section, the survival benefit could have been invoked if the manager had told the people involved they would have to reduce the setup times to less than 3 hours (a number that was picked arbitrarily), or else the entire process would be subcontracted to an outside vendor. This would mean shutting down the process and those employed in the process being either laid off or assigned to other areas within the plant. This is negative pressure, and it can only work if management is perceived as being serious in carrying out this action. There is another version of the survival tactic that managers should recognize. This option promises a lack of risk to personnel involved in the improvement and occurs when management assures employees that they will not lose their positions as a result of any of the improvements. Consider the impact if you, as a manager, were to lay off a group of people who worked on a project that resulted in the elimination of their department. The message would be very clear to the rest of the plant or area: "Don't do it!"

A second approach to WIIFM is that of eliminating *surprises*. Surprises are a fact of life for every production and inventory control manager. While birthday surprises may be pleasant, manufacturing surprises tend to be upsetting. For example, one of our critical pieces of equipment has broken down — surprise! Or, a customer has changed his requirements for an order, and it now has to be moved up in the production schedule — surprise! Surprises complicate life. They force people to reformulate and implement new plans. They distract attention and make production life miserable, for the most part. People will be more willing to participate in a task or carry out changes if they see that the task or changes will eliminate these surprises.

Third is the power of *simplification*. Every person has worked with complexity, either in the form of products or processes. Every person has encountered a situation where the answer is "It depends." Complexity, like a surprise, is very distracting. If the action or task brings with it the promise of simplification (making life easier by eliminating certain options or choices), then people will be more likely to participate. It is a general rule of management that people would rather work with simpler systems and processes than with complex processes or systems. Simplification is one of the reasons that we see so many business process redesign/reengineering projects taking place.

The fourth approach to WIIFM is that of *pride*. People will undertake a task or implement an action if it enables them to do a better job. More people want to take pride in what they do; most people want to be part of a "winner." How else would you explain all the people who suddenly became Cleveland Indian fans when the Indians started to win? (One of the authors picked this example because he has been an Indians fan since the days when they were truly bad.) Linking the task to pride is a powerful motivator.

Finally, there is *direct feedback*. This feedback can take the form of monetary rewards or improved performance on the metrics used to measure performance. This tactic has been used by some firms interested in capturing the knowledge of their departing employees. The management at these firms has paid the people an amount ranging from $1000 to $5000 to share their knowledge (and lessons learned) with the firm. Notice the nature of the relationship — the more valuable my information, the more I am willing to share, the more that I get paid. By using this tactic, we build a direct link between the action or request and the impact on the employee.

Major Lessons Learned

We, as managers, often make the fatal assumption that whatever makes sense and is of benefit to us will be seen in the same way by our employees. This is not always the case. Rather, we must look at tasks from the perspective of the people involved. We must recognize that we get the greatest bang from a plan when the outcomes desired by the plan are seen by the employees as being in line with their own self-interest (hence, the power of WIIFM). When they are, success will inevitably follow.

We saw that there are five approaches that can be used to tap into the power of WIIFM. These include survival, elimination of surprises, simplification, pride, and direct feedback. In each case, there is a direct link between the desired outcome and the self-interest of the people involved. Returning to the scenario presented at the beginning of this section, a WIIFM approach would require that the caller ask himself the question, "How can I make setup reduction attractive to the people involved?" The answers to this question would have then framed that person's approach to reducing setup times.

When a Cost Savings Is a Real Cost Savings

Overview

We now turn our attention to an issue that has been, for some of us, a major source of confusion and, in some cases, embarrassment. This is the issue of costs and cost savings. The confusion that surrounds this topic can best be understood by looking at a situation that involved one of the authors.

You Don't Understand Costs

Some time ago, one of the authors was involved in a project dealing with the in-house manufacturing of a component. The group that had been assigned to the study of this component was very proud of themselves. Over a period of six weeks, they had closely examined a component that was currently made in-house. Everyone agreed that the manufacture of this item was anything but the model of efficiency. The item was fairly large (it was the rail frames for a line of chassis produced by the firm) and was measured and cut by hand. This process was somewhat prone to error, as evidenced by numerous instances of chassis having more holes than were really needed. The flow of the product was problematic. The group found that the rails were marked, then taken outside for storage. Later on, they were brought in for final preparation. During the winter, this meant having to clean snow off the rails, which created a very messy and potentially dangerous work environment. When the group approached several outside frame manufacturers, they discovered that these suppliers could provide rail frames at a lower unit price, in less lead time, and at a higher level of quality. Their recommendation — outsource the manufacture of the rail frames.

Armed with all of this information, the group (with the help of the author) enthusiastically presented their findings to top management. A member of the top management team was the original owner. This person was highly opinionated (to say the least; at one point, he remarked that he thought President Reagan was one of those "damned" liberals). When he had something to say, he said it. In addition to being opinionated, this person was also brilliant — a true intuitive manager and engineer. He could design things off the top of his head that were truly masterpieces. When you asked him how he did it, he could not tell you. His typical response was that it was obvious. Well, this person sat back listening politely. After the group was done, he stood up and asked them why the firm should act on their recommendations. The spokesman replied by noting that outsourcing the rail frames would allow the firm to free up the space currently being taken up by the rail frame department. In addition, the firm would save something like 12% per frame while also being able to reduce the associated warranty costs. Finally (and the trump card), by moving the rail frames out to the suppliers, the firm could save the cost of the four people involved in the rail frame manufacture.

The owner looked at the spokesman and asked if that meant that he would be willing to fire these four people. No, came the answer, these people could be moved to other departments where they could be better used. Well, the owner retorted, we don't have the business to absorb these four displaced workers. The spokesman agreed but noted that if business ever picked up they would not have to hire any more workers. Nonsense, came the reply. What I want, noted the owner, are real cost savings and not potential cost savings. I want money that I can put in my pocket, not the promise of savings. The owner finished by noting that he had taken enough IOUs in the past without having to take another one.

This story taught the author an important lesson — not all cost savings are equal. There are different cost savings, each with its own set of conditions that must be recognized and considered. In general, we can recognize four different categories of cost savings:

- True cost savings
- Opportunity cost savings
- Incremental cost savings
- Myopic cost savings

True Cost Savings

This type of cost savings defines for many the ideal type of cost savings. They are immediate, easily quantifiable, and, therefore, unlikely to generate disagreement. For example, if the supplier of one of our components were to reduce the price per piece by $.50, we would have a true cost savings. This savings is immediate. It is not dependent on the volume of business. It does not have to wait until other benefits are generated so that, in total, there is a significant cost reduction.

True cost savings occur for a number of reasons. They can occur as a result of pass-through savings. The example of our supplier reducing the cost of production and sharing those savings with us is an example of this type of pass-through savings. It could also occur because of a rethinking of the process. We might change the production process so that setup times and the amount of inventory needed are reduced. We could also have a true cost saving due to product redesign. For example, we might find that our customers are not really interested in a feature found in our product. Redesign of this product could eliminate this feature, resulting in a simpler, less expensive product. With true cost savings, our benefits are there. They are not dependent on other conditions.

Opportunity Cost Savings

This type of cost savings is what we saw in the example above and is highly dependent on other conditions to take place. Often, these "other" conditions are defined in terms of volume. In our example, the manpower savings noted by the group would have been gained only if the production volume (reflecting, in turn, increased sales volume) had grown enough to absorb them. Without this increased volume, management has one of two options available. The first is to convert the opportunity cost savings into a true cost saving by terminating the affected employees. Alternatively, management could have kept the employees. Under this option, the cost saving is more of a promise than reality. In short, for an opportunity cost saving to be real, the action must enable management to avoid incurring other costs and expenses.

Incremental Cost Savings

These are cost savings that cannot be realized until they have been accumulated to a sufficient mass. For example, we may have reduced processing time in one area (requiring two operators) by 25%. Because the operation still requires two people, the savings do not really mean much to us (we cannot let go half an operator). These improvements may not have increased overall output, and they may not have affected any other areas; however, this saving is worthwhile. If, over time, we can introduce enough of these changes, the net result may be that we can reduce the staffing requirements for that operation by one operator. Incremental cost savings are like putting money into the bank. We put a little bit away each payday, so little that we do not miss it. Then, at some point in the future, we find ourselves with a fairly nice nest egg.

Myopic Cost Savings

This is a cost savings that is not a cost savings. Assessing the impact of this cost savings requires taking a systems perspective. For example, we are building trucks. The firm is under increased pressure from competition, and every department is encouraged to find and implement any opportunities for savings. One of the departments decides to replace a component in the steering mechanism with another. They are assured that the differences in steering will be minimal. The change is introduced and the cost saving attained. However, not everyone is happy. By changing the nature of the steering mechanism, the handling traits of the truck have been affected, and the customers notice. They bring their trucks back to the shop for repairs. Warranty costs increase and end up being greater than the cost savings initially achieved by replacing the steering mechanism component. With this type of savings, the one department may be better off, but the firm, as a whole, is worse off. In short, a myopic cost savings is a "good news/bad news" situation. A cost savings in one area can create an offsetting increase in costs in another area. The result of such cost savings is chaos.

Major Lessons Learned

- Not all cost savings are equal. Some are true cost savings while others are promises of cost savings.
- There are four major categories of cost savings: true cost savings, opportunity cost savings, incremental cost savings, and myopic cost savings.
- True cost savings are unconditional, immediate, and verifiable.
- Opportunity cost savings require other conditions to take place before they can be realized. Typically, these other conditions are related to changes in production volume.
- Incremental cost savings, by themselves, are not true cost savings; however, over time, these can be accumulated into a sufficient mass to result in a true cost saving.
- Myopic cost savings are those where we save costs in one area, only to lose those savings to problems created by the changes in other departments.
- Knowing the type of cost savings that we have is important so that we can better justify it and explain it to everyone else.

Focus, Urgency, and Time Compression: Understanding the Keys to Success in Implementation

Introduction

We recently received an interesting request from an individual who asked why it was that many of the projects that he was involved in never seemed to reach a conclusion (which brings to mind the old joke about economists — if you were to lay all of the economists of the world end to end, they would never reach a conclusion). We felt that the issues raised in this simple request were interesting enough to warrant being addressed here.

The Scenario

All managers have probably found themselves involved in situations similar to the following (we know that we have). You have just been assigned to a problem-solving task force. The issue facing you is fairly important. You have been asked to determine what can be done to reduce the lead time within a specific segment of the system. At first glance, this seems to be a good assignment. The problem seems to be well defined. You are informed by management that your group will be meeting one hour every week, and that sounds good. Any more time than that would take away from your major work assignments. In addition, you are told to take all of the time that you need. The key is to address and eliminate the problem once and for all. This sounds too good to be true.

You begin the project, and everyone in the group seems to share the same level of enthusiasm. This project will work. However, what you find is that as the weeks pass, you seem to be doing very little on the project. You meet but nothing really seems to be accomplished. The level of enthusiasm seems to be falling faster than the value of Russian stocks. Before long, people both inside and outside of the project seem to realize that it is turning out to look more and more like a failure. The smart people are starting to bail out. Eventually, management disbands the team, and this dream project seems to have turned into the project from hell. Now, there you sit wondering whatever happened to this dream project.

The Problem

The reality is that this dream project was really doomed from the start. The reason is that it lacked three major attributes — focus, urgency, and time compression. Whenever these three traits are missing, failure is sure to follow. Why are these three traits so important? Before we answer that, let's begin by looking at what went wrong with the project.

First, we see that the project really had no sense of urgency. There was no real deadline. Instead, the project team was simply asked to tell management when they were done. In addition, the team was doomed by the schedule. They were asked to meet one hour a week. With most teams, meeting once a week for only one hour is almost guaranteed to fail. Why? Have you ever tracked the amount of time actually spent working on the task at hand in a one-hour meeting? Let's take a look at how time is used in one of these meetings.

First, we never really begin on time. It takes about 5 to 10 minutes before everyone arrives for the meeting. This means that we now have only 50 minutes left. We next have about 5 minutes of socialization and gossip, leaving us with 45 minutes. Next, we spend about 10 to 15 minutes rehashing what was discussed at the last meeting. This leaves us with about 35 minutes. Then, we review our assignments from the last meeting; again, this takes about 10 minutes (in most cases, most of the assignments have not been done because other more critical tasks interrupted them; there always seems to be a good reason for the lack of work). This leaves us with about 25 minutes. Then, we have to think about the activities for the next meeting, which takes about 5 minutes. Finally, we seem to break up early because some members of the team are always being called out of the meeting 5 to 10 minutes early. This leaves us with between 10 to 15 minutes in which we are actually doing anything. Is it any wonder that we get so little done?

To overcome the problems inherent in such a schedule, we need to recognize the necessity of the three attributes previously identified. First, let us begin with *focus*. Focus means that from the outset we have clearly delineated the task to be studied. We have bounded the problem area in terms of span (how much of the system we will study) and conditions (specific conditions under which we are to study the problem). Once we have established a bound on the problem, we do not go beyond it. If we detect a problem that lies outside of the bounds, we do not ignore it; rather, we use the vehicle of the "action list" to record the problem. That is, we write down the problem, the reason that it is important, the implications of not dealing with it, and the type of actions that must be considered to deal with it. We leave the items on this action list for others to focus on. For ourselves, we deal only with those items within the bounds previously set down. At first glance, this seems like an unnecessarily stringent limitation, but there is logic to it. If we do not set down such a bound, then the problem seems to grow and become bigger over time. When problems grow, they never seem to be resolved. There is always one more factor or scenario that should be considered.

A second issue is that of *urgency*. Urgency is the idea that people feel the problem that they are dealing with is important and they must address it immediately. The schedule in our example was not consistent with the notion of urgency. To establish urgency, we must demand that the participants work on the project full time for a period of time. Their assignment is their job. This is in contrast to the approach we saw in our example, where the assignment was something that the group members fit into their work schedules as they saw fit. With this latter approach, the message to the group members was quite clear — their daily jobs were the real priority. This is not the message that we want to send.

By the way, urgency has an interesting corollary associated with it. If we assign people to a problem team on a full-time basis, then we must have others available to cover their daily responsibilities while they are serving on the team; otherwise, we create a situation where the project team members know that they can expect a pile of work to be waiting for them upon their return. These people have helped the company address a major problem, and their reward is to end up further behind in their workloads. This is hardly the proper reward to give people.

Finally, there is *time compression*. For most projects to succeed, we have to set a specific deadline. The deadline should not be too far into the future; for most projects, where people are assigned to them on a full-time basis, the task should be completed with a high degree of success within two to five days. Also, the deadline should be a drop-dead one; that is, management should be prepared not to offer any extensions. If the team members ask for an extension because otherwise there might not be anything to report, then management should be prepared to say that if that is the case, then that is what the group members must report to the executive team supervising the project. Such an approach gets people's attention because it conveys the seriousness with which management views the timely attainment of results. With these three elements in place, we greatly increase our chances for success.

Hey, Vendor, There's a Paradigm in my Software!

Overview

At the heart of every manufacturing software package is a framework or paradigm. When we buy such a package, we are buying more than the medium on which the program was delivered; rather, we are buying the paradigm or approach taken. If this paradigm does not fit with the way that we do business or the way that our firm or system is run, then the results can be potentially very costly and dangerous.

The Manufacturing Software Decision

For nearly every manager (especially at the mid to upper levels), one of the most important decisions that they must make involves the selection of the most appropriate manufacturing package for their firm. This is an important decision for several reasons.

First, the proper package can greatly enhance the competitive capabilities of their system. It can make the good system better. It can simplify decision-making by its personnel. It can also generate better decisions and plans in shorter lead times. It can enable the firm to replan quickly and meet the demands of a changing environment in an efficient and effective manner.

Second, there are the costs associated with this decision (which are themselves quite significant). Included in these costs are the costs of identifying potential packages, evaluating these packages, visiting clients using the packages being considered, the acquisition of the packages themselves, employee training, consulting, hardware acquisition (and upgrading), implementation, debugging, and maintenance (to name just a few of the costs). Third, there are the costs of changeover (if we find that we picked the wrong package).

Finally, we must recognize that the probability of failure is very high. Traditionally, in the MRP/MRPII field, the failure rate has ranged between 80 and 95%. There is no reason to suspect that this failure rate has improved, as we have moved into the more complex environment of Enterprise Resource Planning (ERP). It is this last reason on which we will focus. If you think about it, this high failure rate makes the manufacturing software selection decision a very unattractive one. It is a bet where the odds are stacked severely against the manager; however, it is a bet that we must make if we are to succeed. Many consultants and writers have identified possible reasons for this very high failure rate (inadequate training, poor manual systems, incomplete or inaccurate information, to name a few), but there is yet another factor overlooked by many managers which must always be considered. Every manufacturing package out on the market embodies within it a paradigm or framework.

Paradigms play an important role in the life of a manager. They tell us what we should focus on; they also identify those areas that we can overlook or ignore. They identify opportunities for performance. Paradigms are important in manufacturing because of the inherent complexity of our task. Every manager knows that what works in one setting may not necessarily work in another. The same is true when it comes to manufacturing software (especially major packages such as MRPII/ERP systems). It is impossible to develop a package that is totally comprehensive. Such a package would extract its own penalties in terms of size, operating system requirements, and operating speed, and no software vendor has ever encountered every possible setting. As a result, developing a manufacturing software package becomes a very different task when compared to the task of developing a word processing package. There is a strong commonality of tasks and activities that simplifies the development of a word processing package. We have to format words, paragraphs, and pages; we have to be able to edit our document and have the package automatically rearrange the remaining text; we would like to be able to seamlessly insert graphics and spreadsheet material. Such commonality is lacking when it comes to manufacturing.

Given this inherent complexity, one way that vendors try to simplify their tasks is to draw on paradigms. Sometimes these paradigms are based on industrial settings; in other cases, they are based on "theoretical" approaches to the management of manufacturing; in still other cases, the packages are built around certain algorithms or procedures. As an example of the first paradigm, we know of one package that was developed for managing operations within a European petrochemical plant. Packages built on approaches such as the Theory of Constraints or Flow Management represent the second category. Finally, we see the third category in packages that incorporate generic algorithms, simulations, or some form of mixed/dynamic or linear programming.

Why is it important that we understand the nature of the algorithm driving the specific manufacturing software package? The answer is that when we buy the package, we buy the algorithm. When we buy the algorithm, we accept either implicitly or explicitly the assumptions underlying that algorithm. If these algorithms are not appropriate to our setting, we may find ourselves encountering unexpected problems. These problems may reduce the speed with which we implement the software package; they may also require that we contract for additional programming (to bring the package into closer agreement with our operating environment). All of these effects effectively lengthen the implementation lead time and increase the overall implementation costs. They also increase user frustration. In aggregate, these are effects not conducive to the successful implementation of any software package.

To understand the impact of these effects, consider the following situation described to us by a professor who spends a great deal of time working in Europe. A well-known vendor had sold a package to a major aircraft manufacturer. The manufacturing control package had been initially developed for use in a European petrochemical plant. In this environment, work flows are linear/ sequential. In addition, a major concern is that of scheduling and managing the flow of work to the first or gateway work station. Also, the environment allowed the developers of the package not to have to worry about user discipline. Serial tracking was also not a major consideration. The buyer received assurances that this package would meet or exceed all of their expectations; however, when they started the implementation process, management noticed that things were not going as well as they had expected. Ultimately, the firm had to contract with the software vendor for additional programming. The result was a large, unexpected increase in costs and lead time.

At times, the influence of paradigms can be subtle. For example, one firm, which dealt with a lot of low usage but very expensive parts, found that its inventory levels were suddenly increasing after the implementation of a manufacturing package. What they found, after a great deal of time and effort (most of which was spent plowing through the code), was that the safety stock calculations in the package were based on calculations that rounded up all safety stocks to the nearest integer. For a part where the average annual demand was 1/10th of a part per year, the result was a major increase in costs.

So What?

We have introduced yet another dimension that must be considered when evaluating potential manufacturing software packages, and now we will explain why this discussion is so important. Far too often, when we approach the process of picking a software package, we focus on issues such as features and costs. We are interested in identifying that package having the set of software features that we think are appropriate. Far too often, we try to be comprehensive in our selection of packages. We look for that package that has all of the features we might even considering using (even if we never use them). Typically, we ignore the issue of the underlying paradigms or frameworks. By introducing the issue of paradigms, we are effectively adding certain new questions to our software selection process. We are now asking:

- In what type of manufacturing setting was this software package developed? What were the major traits of this environment?
- What type of users were present when this package was developed (how well disciplined were they)?
- What data accuracy and/or completeness were present when this package was developed?

As we noted earlier, every manager who undertakes a software implementation is faced with fairly long odds against its success. We must do everything that we can to ensure that we keep the odds in our favor. By focusing on the underlying paradigms, we recognize another dimension that must be considered when evaluating packages.

Postscript: Finding a Beginning in the End

By reading this book, you have been introduced to many of the basics of manufacturing excellence; however, simply knowing these basics does not necessarily ensure manufacturing excellence. Make no mistake about it — manufacturing excellence cannot be attained without a thorough mastery of these basics. These basics are similar to the tools used by an expert carpenter, mechanic, or craftsman; for these experts to produce their masterpieces, they must have a thorough knowledge of what each tool is, what it does, how to use it, and, most importantly, when that tool is appropriate and when it is inappropriate. You have started to develop that same type of knowledge base by reading this book. You have started to build a strong foundation on which manufacturing excellence can be built. The next step is to move from mastery of the tools to manufacturing excellence. This movement is a process that consists of a series of well-defined steps. Each step must be thoroughly dealt with before you can proceed to the next step.

1. Make a list of the basics discussed in this book that you view as being highly appropriate in your organization or process. For each of these basics, identify the reasons why it is appropriate. Also, evaluate the ease

with which you can quickly, easily, at low cost, and without much disruption implement these basics. If you can identify a subset of basics that you can implement quickly and at a low cost, then implement them. These are the "quick hits" — they generate the benefits without the costs. They also enable you to resolve those problems that have probably plagued and annoyed others in the organization. If successful, then they can help you, as the manager, gain a strong sense of credibility with those who have lived with these problems.

2. Identify the desired outcomes of the area or the process that you are responsible for. These outcomes should be described in as specific of terms as possible. A stockroom, for example, is responsible for more than storing items. In most cases, the stockroom is responsible for three outcomes: (a) ensure that the number of items on the shelves correspond, within the tolerance, with the numbers found in the computer databases; (b) provide timely access to the items in the stockroom when required by the users; and (c) be asset managers (identify the slow movers or the obsolete items and take the appropriate actions aimed at getting them out of the system). These outcomes should be described in as much detail as possible because, as previously noted, they will serve as the basis for generating metrics.

3. Once you have identified the desired outcomes, assess the extent to which the processes or activities are able to achieve these outcomes. If the level of performance is unacceptable, then determine the potential reasons for this poor level of performance. Using these reasons as a guide, identify areas and processes that you should focus on. This process of identification should be done with the assistance of those people who are involved in the process or who have a good working knowledge of the processes and the areas. These are the real experts, and they are also the people who will have to use any potential solutions once they have been developed and implemented. When drawing up this list, you should consider three criteria: (a) frequency (how often this process or activity seems to be responsible for the observed difficulties), (b) severity (how bad a problem is when it occurs), and (c) ease of implementation (how easily we can implement or carry out the changes or affect the process/activity). An ideal candidate should be one that has a high degree of frequency and high severity (when a problem occurs) and is relatively easy to address. The result of this step should be a prioritized list of areas to examine beginning with the most critical and ending with the trivial. This prioritized list is critical because it helps to direct activities. That is, you begin at the top and move down through the list. By the time you get to the end of the list, the problems should be eliminated.

4. Begin at the top of this list. Having identified an area to focus on, be sure that you bound the processes associated with this area tightly.

You do not want to let people move beyond this area. If they identify an area that should be investigated, then put it down on an "action list." An action list, as previously noted, identifies areas that management should look at in the future. Such lists help shape future improvement activities.

5. With the area identified and bounded, identify the processes associated with this area and document them as they are. For example, if you choose to look at an inventory accuracy problem, look at the steps that you use to move a part through your inventory system. If you have a bill of materials accuracy problem that you want to resolve, document the processes that created the bill of materials. It is important to reemphasize a point previously noted in this book — you must document the process as it actually exists (not what you think exists). To do this requires that you get out in the system and actually study the process. You cannot correct an inventory accuracy problem by staying in your office.

6. Further refine the areas that you want to focus on. Within the process, pick those areas that you feel are most responsible for the observed problems and difficulties. Identify opportunities for improvements. These opportunities should come from your study of the operation of the actual process. Identify those actions and changes that should address these opportunities, and implement them immediately. Your role as management should be to remove the obstacles rather than putting more in place. For example, if you choose to look at an inventory accuracy problem, look at the steps taken to move a part through your inventory system. If you have a bill of materials accuracy problem to resolve, document the processes that created the bill of materials.

7. Implement those changes in a manual mode so that you can see what is really going on. By monitoring only one part, you can make additional changes as you learn the dynamics of your operation, and, if your ideas did not work as well as you had hoped, it is not a major problem to make changes again. Furthermore, you have learned something important — you have learned what doesn't work (and why not).

8. When carrying out changes, remember the lessons of urgency and time compression. Actions must be carried out within a limited time period; otherwise, we run the risk of losing focus.

9. When you have identified all of the changes necessary to improve operations and their underlying processes (consistent with the desired outcomes as discussed in item 2), document the changes and make sure that this new documented process is the process that everyone must follow. Included in this step are issues of training (leading users through the new process) and education (explaining to managers and others the need for the changes).

10. When you are satisfied with the changes, then and only then can you proceed to the next problem area. Then expand on to the next, and so on, until all areas have been investigated and appropriate changes implemented. In the inventory example, move on to the next part after changes to the first part have proven effective.

11. Communicate the changes to others in the firm. This is especially important when dealing with upper management. Their support and commitment are needed if you want to see your solutions supported and the program of ongoing improvement continued. This communication is critical because, as we improve the processes, we are also generating new capabilities for the firm. These new capabilities can affect how the firm competes in the marketplace. As a result, these improvements may ultimately have strategic benefits, an issue of vital concern to top management.

12. Communicate the action list with management. The action list, developed from this process, identifies other opportunities for improvement. Management should be involved in evaluation of this list and in the selection of other candidates for future improvement projects.

There is far more to the task of making the transition from mastery of the basics to manufacturing excellence than can be covered in this book. We have only uncovered the tip of the iceberg, the most visible issues. The tip of an iceberg, as we know from research, is only 10% of the total mass of the iceberg. We have focused attention on the most obvious aspects of the iceberg; however, we know that when we attack the tip of the iceberg and reduce it, the iceberg gets lighter. As it gets lighter, the iceberg rises and new problems are exposed. As these newer problems arise, they tend to be somewhat different from the ones that we have just addressed. As a result, new and different approaches are needed. These new and different approaches will be discussed in our next book, which will take you below the surface and provide the basics necessary to attain the next major level of improvement. With every step we take in this process, we get closer to the goal of manufacturing excellence.

In the meantime, if you have any comments, suggestions, or questions, please do not hesitate to contact either one of the authors. Steve Melnyk can be reached at Michigan State at 517/353-6381 or on his e-mail at **melnyk@ msu.edu**. Chris Christensen can be reached at the University of Wisconsin at 608/441-7326 or on his e-mail at **cchristensen@bus.wisc.edu**. Until then, good luck on your journey. The results are well worth the efforts.

Index